THE WAR
AGAINST
AMERICA

THE WAR AGAINST AMERICA

SADDAM HUSSEIN AND THE WORLD TRADE CENTER ATTACKS

A STUDY OF REVENGE

LAURIE MYLROIE

SECOND REVISED EDITION

Foreword by Former CIA Director R. James Woolsey

Previously published as *Study of Revenge*

ReganBooks
An Imprint of HarperCollins*Publishers*

SECOND REVISED EDITION

Library of Congress Cataloging-in-Publication Data has been applied for.

ISBN 0-06-009771-X

01   02   03   04   05   COG/RRD   10   9   8   7   6   5   4   3   2   1

*In memory of my father, Robert Mylroie, and
for my mother, Augusta,
without whose enduring love
this book could not have been written*

I ask myself, what type of person shows no regard for human life and would bomb the most populated building in the world?

> —Edward Smith, who lost his wife, Monica, pregnant with their first child, Eddie, in the World Trade Center bombing, following the May 24, 1994, sentencing of the first four defendants in the bombing

What though the field be lost?
All is not lost; th' unconquerable Will,
And study of revenge, immortal hate,
And courage never to submit or yield;
And what is else not to be overcome?

John Milton, *Paradise Lost*, Bk. I, 105–09

Contents

CHARTS

Foreword

The catastrophe produced by the assault against the United States on the morning of September 11, 2001, brought home the horror of America's vulnerability to terrorism. Seared into the nation's consciousness are the terrible images: New York's tallest tower aflame, black smoke billowing from a gaping hole where a passenger plane has hit the building, while a second jetliner speeds, incredibly, full throttle into the other tower and bursts into a giant fireball. The Pentagon is hit as well, and then, less than an hour later, a stunned nation watches as the twin towers of the World Trade Center crumble to the ground.

In the aftermath of these attacks, the urgent question for all of us has been, Who was responsible for this atrocity? Our government faces very difficult decisions about how to respond; and, of course, the response must depend on a determination of who was responsible. We naturally look to the official investigation for the answers.

Few hard facts are available publicly, however, as of today. Usama bin Laden appears to have been involved, but a key question remains: Did he and his network act alone? Investigating a major terrorist attack is necessarily a long and painstaking process, and this investigation is particularly difficult, given the magnitude of the assault and the steps taken by the attackers to hide their tracks and give false leads. But precisely because a trail has been left that points so obviously to bin Laden, the nagging question for those of suspicious mind is whether there may have been a senior partner hiding in the shadows, carefully concealing its role from investigators and now encouraging us to look elsewhere.

Have we, by any chance, been here before?

The War Against America, by Laurie Mylroie, is highly relevant now for several reasons. Mylroie has meticulously examined the evidence surrounding the 1993 conspiracy to bomb the World Trade Center—a conspiracy that failed in its key aim, to bring down the twin towers. Her overriding concern, evident throughout these pages, was that the 1993 Trade Center bombing presaged further attempts at terrorism on a massive scale on American soil. She points clearly to the limitations of a strictly judicial inquiry into a case involving international terrorism. Most dramatically, she shows that the question of the identity of the plot's mastermind, Ramzi Yousef—never to this date satisfactorily resolved—has very serious implications for our understanding of whether there was a government behind the first attack on the Trade Center.

In our efforts to unravel the conspiracy behind the 2001 attack, we need to reexamine the evidence from the 1993 attack, which was never (as this book demonstrates) satisfactorily explained. Mylroie argues compellingly that the lack of official attention to the international context of the 1993 attack, and particularly the failure to address the question of state sponsorship of the terrorists who carried it out, helped leave America open to further terrorism.

The War Against America presents a careful and lucid analysis of the 1993 attack. The central argument of this brilliant and brave book is that the Iraqi government was key in the planning and implementation of that attack and, more specifically, that Ramzi Yousef was himself an Iraqi intelligence agent. This was in fact the tentative conclusion reached by the veteran counterterrorism expert Jim Fox, then director of the New York Field Office of the FBI, who led the investigation until early 1994.

Instead of directly addressing Fox's strong suspicions that Iraq was behind the 1993 bombing, however, the Clinton administration chose to treat this act of terrorism purely as a law enforcement issue. Emphasis was placed on arresting individuals, trying them in court, and securing convictions. The conventional wisdom (promoted by the administration) that soon emerged regarded this and indeed most any terrorist attack on the United States as the likely product of a "loose network" of folks who just somehow got together or, later, as the exclusive work of the elusive network-meister bin Laden. The possibility of state sponsorship, especially by Iraq, was in effect not on the official agenda.

Readers can determine for themselves whether the blinkers that narrowed the field of vision of the Clinton administration and focused it on networks, not states, were created solely by an objective analysis of the facts as the administration knew them at the time, or were the result of something else. The Clinton White House had a propensity to start with the impression it wanted to create and then work backward to the policy it followed and even the facts it sought. If you start from the proposition that you don't want the bad news that would result from a clear confrontation with a state, the probable resulting hostilities, and the likely casualties that would result, then putting on blinkers that leave state actions outside your field of view will help serve your purpose.

Whatever the administration's reason for choosing this fatally flawed approach to countering terrorism, once it was adopted two aspects of the U.S. legal system cut in to constrict the government's ability to look for evidence of state sponsorship for the duration of the criminal investigation and trials.

First, a prosecutor's team is not the right institution to use to look for an overall assessment of whether there is state sponsorship of a terrorist act. Turning the issue entirely over to the prosecutors almost guarantees a heavily circumscribed approach, and not through any fault of the prosecutors themselves. Indeed, the better the prosecutors are, the more likely they are to focus like a laser on proving that the people they can get their hands on have committed the elements of the crime set out by the law—not on a general search for background useful to some other part of the government. Much like a tactical military commander in the field, a prosecutor preparing to put a case before a jury looks for the simplest and clearest approach that will win. Trials are not general searches for truth and insight. They are legally circumscribed fights, a kind of juris-prudential trial by combat. For a prosecutor before a jury, complexity is the enemy and simplicity is a friend, just as is the case for a Marine company commander planning to take a hill. It makes as much sense to expect a prosecutorial team to make an overall assessment of state sponsorship of a terrorist event as it does to ask a Marine captain, in the midst of giving tacti-cal orders about fire and maneuver to his platoons and squads, to pause for a bit and reflect on the politico-military connections and alliances of the enemy whose troops he is facing.

The prosecutors in the 1993 World Trade Center bombing case and the second New York conspiracy to bomb several tun-nels and buildings in New York had a simple, coherent theory: that both the Trade Center bombing and the second plot, foiled by an FBI undercover operation, were part of one overall con-spiracy inspired by Shaykh Omar ("the blind Shaykh"). This theory was not at odds with the facts, it had the virtue of sim-plicity for the jury, and it may well have been the truth and nothing but the truth.

But there is a chance that it was, unknown to the prosecu-tors, not the whole truth.

A second feature of the U.S. judicial system kept the intelli-gence community, the National Security Council, and indeed anyone who might have been able to focus on the issue of possible state involvement from looking at most of the relevant information until many months after the bombing.

Grand jury secrecy, as codified under Rule 6(e) of the Federal Rules of Criminal Procedure, severely restricts the flow of information to the rest of the government from investigations such as this one, where the facts should compel the government as a whole to look at national security issues as well as the needs of law enforcement. The rule permits grand jury material to be shared with other law enforcement authorities in certain cases—for example, to help solve or prosecute other cases—but there is no exception to permit intelligence officers to see such material in order to make an assessment about possible foreign government support for a terrorist act.

Consequently, no one other than those immediately involved in the 1993 World Trade Center bombing prosecution in the Justice Department, the U.S. Attorney's Office in New York, and the FBI had access to the evidence being collected until some of it was presented in court at the trial. The telephone, passport, and other records introduced into evidence were extensive, although only a small portion of this material was addressed during the trial. Following the trial, only one person, as far as we know, had the wit and the diligence to plow through all of these materials and, in the process, gleaned from them the possibility of Iraqi government involvement: the author of this book.

Grand jury secrecy generally has much to commend it. Investigations often need to be secret in order to be successful. Secrecy is also needed to protect the privacy rights of subjects and targets of investigations, who should not be publicly subjected to false accusations, and much material brought before grand juries does not pass the test of what is required for an indictment or what may be admitted at trial.

Yet it is difficult to deal effectively with terrorist activity if, by the act of setting foot on American soil, a terrorist can ensure that the evidence of his actions will be shielded from our national security agencies as the legal process runs its course. Some way needs to be found to provide such information to those parts of the U.S. government that can assess foreign government involvement in a terrorist act. These are, after all, the government agencies with the most expertise in international affairs and the greatest ability to find out whether there is any state sponsor. If the intelligence analysts do their jobs properly, they may, for example, be able to detect a "false flag" operation—one in which a government ultimately behind an operation takes care to shield its involvement by casting suspicion on a third party. The third party may well not be innocent; indeed for some such operations, the more it is involved the better. What is key to a government seeking to hide its hand is that the wizard pulling the levers behind the curtain not be disclosed. It would have been a classic false flag operation if Iraq in fact had sponsored the 1993 World Trade Center bombing and left a handful of Muslim extremists behind to be arrested and take the full blame.

If a deception serves its intended purpose, the state sponsor of the terrorist attack has incurred little cost and no consequences.

In such a case, not only does a successful prosecution of the perpetrators do nothing to prevent future terrorist attacks, it in effect encourages the sponsoring state to carry out further terrorist attacks as false flag operations.

And that, Mylroie cogently argues, is precisely what has happened since 1993. A year before September 11, 2001, Mylroie wrote:

> The Clinton administration has consistently turned a blind eye to the clear and obvious dangers that Saddam poses. If America's leadership continues to deal with Saddam in that fashion, we must be prepared to see further acts of violence that are more successful, more brutal, and more devastating.

Those lines may now have taken on a tragic cast of prescience. If time proves that Laurie Mylroie is right about what happened in 1993, the truth may yet help us, belatedly, to find our real enemy and defeat him. If so, try to think of a living American to whom you would owe more.

R. JAMES WOOLSEY
Washington, D.C.
September 27, 2001

Acknowledgments

This book has been in the making for some years and over that time, a number of people have shown a special concern for it. Their sustained interest and efforts on its behalf constitute a debt, the extent and nature of which I cannot adequately express.

In particular, I would like to thank two friends who have supported this project from its inception. One is Clare Wolfowitz. She fundamentally shaped this book, infusing it with her wisdom, prudence, and very considerable talent for expression. Her input was an invaluable gift from an extraordinarily generous heart.

The other friend to whom I am especially indebted is John Hannah. His clear eye and sober judgment have been a source of guidance and support over many years. That includes the matters dealt with in this book and also goes far beyond them.

Alan Berger has also lent his assistance to this project since it began. In fact, when he first read this work, it was an article grown too long. As a former professor of English, he recognized it as a book and encouraged me to write it up as such. Since then, he has provided many extremely insightful comments. Paul Wolfowitz was kind enough to listen to this work presented orally and then later to read the manuscript. At critical times, he provided crucial support for a project that is inherently difficult. I am extremely grateful to both, as I am to Lewis (Scooter) Libby for his timely and generous assistance.

This book brings together several diverse fields, and I am very much obliged to individuals in those fields who regularly shared their expertise with me.

They include several people in New York law enforcement, above all Jim Fox, who was director of New York FBI at the time of the World Trade Center bombing and led the FBI investigation in New York. I first met him in the summer of 1994, when he was retired from the FBI and worked at Mutual of America. He had believed that Iraq was behind the bombing and he was extremely helpful in the research for this book. Law enforcement is a closed world, but Fox was very generous to this outsider. As his personal assistant at Mutual of America, Maria Aleman, would say, "He always had time for a good cause." Fox liked to tell people about his father, the bus driver—and his father's advice, "Do the work you love and you'll never have to work a day in your life." That's how Fox felt about the FBI. Sadly, he passed away suddenly, in May 1997. I am indebted to him and to Maria.

I am also grateful to Gil Childers, who has since left government, but who led the first World Trade Center bombing prosecution and who was to lead the prosecution of Ramzi Yousef, if he had not been tapped for another duty a few months before Yousef's trial began. Childers read an early version of this work and was kind enough to answer my questions and share what information circumstances allowed. Several others in New York law enforcement are best thanked anonymously.

This work also covers intelligence matters and Vincent Cannistraro, former chief of counterterrorism operations for the CIA's Counterterrorism Center, has shared his insights and expert advice over the years. Warren Marik, retired from the CIA and presently president of Information for Democracy, has also been an invaluable guide.

Rolf Ekeus, former UNSCOM chairman, and Ewen Buchanan, UNSCOM spokesman, very kindly helped keep me informed of what was known (and unknown) about Iraq's unconventional weapons programs. Richard Speier, retired from the Pentagon, was unfailingly ready to answer further questions about technical issues related to those programs.

This work also owes much to the keen insights of Francis

Brooke into Washington politics and the formulation of U.S. policy toward Iraq. Sharon Brooke read and provided helpful comments on this work. The Brookes are warm and generous friends, with whom over the years I have shared the frustrating experience of watching what seemed to be America's overwhelming victory over Iraq erode into the precarious situation described in this book.

An article based on the material in this book, "The World Trade Center Bomb: Who Is Ramzi Yousef? Why It Matters," appeared in *The National Interest* (Winter 1995–1996). I am most obliged to *National Interest* editor Owen Harries for providing an early forum for this work, as I am to the Bradley Foundation, which provided partial support for this project.

But the greatest debt of all is to the American Enterprise Institute and especially to AEI president Chris DeMuth, who gave the book a favorable reading, as did Seth Cropsey, Michael Ledeen, and David Wurmser. John Bolton provided additional assistance. The support of the outgoing director of AEI Press, Jim Morris, was very much appreciated, as is that of his successor, Monty Brown. Thanks are also due Virginia Bryant, Kenneth Krattenmaker, and my editor, Cheryl Weissman, who is acknowledged with extra special appreciation for her work on this book. To all of them, I am extremely grateful.

Cast of Characters

SHAYKH OMAR ABDUL RAHMAN
Blind Egyptian cleric, convicted of directing a seditious conspiracy against the U.S. government, a war of urban terrorism.

MAHMUD ABU HALIMA
Egyptian Muslim extremist, friend of El Sayyid Nosair. Nosair's intended getaway driver after Meir Kahane's assassination. Subsequently convicted in the World Trade Center bombing.

MUHAMMAD ABU HALIMA
Younger brother of Mahmud. Originally charged in second bombing conspiracy, but those charges were dropped and he was tried and convicted for helping his brother flee the country after the Trade Center bombing.

KADRI ABU BAKR
Muhammad Salameh's uncle and once a member of the Palestine Liberation Organization's "Western Sector" in Baghdad.

AHMAD AJAJ
Palestinian, entered the United States with the World Trade Center bomb mastermind and was convicted in the World Trade Center bombing.

SIDDIG IBRAHIM SIDDIG ALI
Sudanese, organized the second bombing conspiracy, the attempted bombing of the United Nations and three other New York targets. Turned state's evidence in January 1995.

NIDAL AYYAD
Palestinian émigré to America, close friend of Muhammad Salameh, and convicted in the World Trade Center bombing.

IBRAHIM EL GABROWNY
Older cousin of El Sayyid Nosair. Arrested after World Trade Center bombing for resisting arrest and convicted in the second trial, with Shaykh Omar et al.

EYAD ISMAIL
Palestinian, drove the van that carried the World Trade Center bomb but believed that there was shampoo in the van.

ABDUL HAKIM MURAD
Baluch, convicted with Ramzi Yousef in the plot to bomb a dozen U.S. airplanes.

EL SAYYID NOSAIR
Egyptian, charged in the November 1990 murder of Meir Kahane. He was acquitted of murder but convicted on lesser, related charges.

MUHAMMAD SALAMEH
Palestinian, participated in the World Trade Center bombing. He was arrested as he returned to the Ryder rental agency to pick up the deposit on the van he had rented to carry the bomb.

EMAD SALEM
Egyptian informant to the FBI, working for Egyptian intelligence at the same time that he was working for the FBI.

WALY KHAN AMIN SHAH
Uzbek, convicted with Ramzi Yousef in the plot to bomb a dozen U.S. airplanes.

ABDUL RAHMAN YASIN
Iraqi, born in America, came back to the United States in September 1992. He helped mix chemicals for the bomb, then fled to Baghdad. The sole remaining fugitive.

MUSAB YASIN
Older brother of Abdul Rahman.

RAMZI AHMED YOUSEF/ABDUL BASIT MAHMOOD ABDUL KARIM
The man most responsible for building the World Trade Center bomb. He arrived in the United States on September 1, 1992, on an Iraqi passport under the name "Ramzi Yousef" and fled on the night of February 26, 1993, on a Pakistani passport as "Abdul Basit." He was arrested in Pakistan and returned to the United States February 8, 1995.

Timeline: 1992–1993

EARLY JUNE 1992: El Gabrowny recruits Salameh into the original bombing conspiracy: Nosair's pipe bombs.

JUNE 10: Salameh makes the first of 40 calls to his uncle in Baghdad, Kadri Abu Bakr, formerly no. 2 in the PLO's "western sector."

JUNE 22: Abdul Rahman Yasin applies for a U.S. passport.

SEPTEMBER 1: Ramzi Yousef arrives at JFK Airport with Ahmad Ajaj, whom he met in Islamabad. Ajaj is detained; Yousef goes to 34 Kensington Avenue, Jersey City.

CIRCA SEPTEMBER 1: Abdul Rahman Yasin comes to America and lives with his brother, Musab.

OCTOBER: Yousef and Salameh move out, to their own apartment.

NOVEMBER 2: George Bush loses the presidential election.

NOVEMBER 3: Saddam goes to Ramadi and fires his pistol in the air, saying, "The mother of battles continues and will continue."

NOVEMBER 18: The first call related to the purchase of equipment for making the bomb appears on Ramzi Yousef's telephone bill.

NOVEMBER 30: Salameh rents the storage locker used to store the chemicals.

DECEMBER: Ramzi Yousef makes the calls to secure his escape route.

DECEMBER 31: Ramzi Yousef goes to the Pakistani consulate to ask for the passport on which he will flee.

JANUARY 1, 1993: Salameh and Yousef move to the apartment in which they build the bomb, set back from the street.

UNKNOWN DATE: A mystery man arrives to supervise the bombing in its last stages.

FEBRUARY 26: At 12:17 P.M., bomb explodes on the B-2 level of the World Trade Center parking garage, next to the support columns of the North tower. Bomb mastermind Ramzi Yousef flees in the evening.

MARCH 4: Salameh is arrested; Abdul Rahman Yasin leaves the next day.

1

Introduction

Prior to September 11, 2001, Americans generally considered themselves safe from foreign attack. Two oceans, friendly neighbors, and a decade of peace and prosperity contributed to an unusual sense of security and well-being shared by most Americans. That illusion has been shattered.

We are all very familiar with the terrible, bloody events of September 11. At 8:45 A.M., Mohammed Atta, an Islamic militant born in Egypt, crashed a Boeing 767 into the New York World Trade Center's North Tower. At 9:06 A.M., Marwan al-Shehhi, Atta's close friend and constant companion, flew another 767 into the South Tower. At 9:40 A.M., Hani Hanjour, a Saudi, drove an airplane into the Pentagon. And at 10:37 A.M., United Flight #93 crashed in the Pennsylvania countryside, when a heroic group of passengers prevented the pilot, a Lebanese named Ziad Jarrah, from attacking his intended target.

It quickly became clear that Usama bin Ladin and his group, al-Qaeda, were involved in the planning of these attacks. Two of the hijackers, suspected associates of bin Ladin, were on a U.S. "watch list," supposedly prohibited from entering the country. But the critical question remains: Who was ultimately responsible for providing the direction, expertise, and logistical support for the attack? Could this complex plan have been executed by al-Qaeda alone, or did some other, more powerful entity under-write the attacks?

Mohammed Atta has been called the chief organizer of the September 11 assault. He was the conspirator who received a large cash transfer from abroad, and he had gone to some effort to meet with senior Iraqi officials, once immediately before his first, fateful trip to the United States and again five months prior to the attack. While the initial blame has been pinned almost exclusively on the al-Qaeda network, there are many clues that point to a more powerful co-conspirator: the government of Iraq. Indeed, the history of Saddam Hussein's involvement offers an eye-opening blueprint to the September 11 attacks, in the form of the first assault on the World Trade Center—the bombing of 1993. The story of that bombing, and of its convicted perpetrator Ramzi Yousef, reads today as an early warning of the far more horrific events of September 2001.

This book offers a detailed reexamination of the facts surrounding the first attack on the World Trade Center. It presents compelling evidence that the individuals involved did not act alone. And, in the process, it exposes the FBI in the "mistake of the century," as one distinguished former U.S. ambassador to the Middle East described it.

And, ultimately, *The War Against America* argues that the first assault on the World Trade Center did indeed have state sponsorship—from Iraq. It presents the case that Saddam Hussein is the single greatest terrorist threat to America. And it concludes that his campaign against the allies of the Gulf War continues, almost undetected, to this day.

How did Washington fail to see that Iraq was behind the first World Trade Center bombing? How did Saddam escape blame? The answers to these questions can be found in our changing definition of terrorism.

Before 1993, the official view in Washington was that major attacks on American targets were, almost invariably, state-sponsored. After any major bombing attack, it was assumed that a terrorist state was responsible. For all practical purposes, that meant Libya, Iran, Iraq, or Syria. In those days, terrorism had an address.

But over the past eight years—starting with the first attack on the World Trade Center on February 26, 1993—a new explanation for terrorism has gained widespread acceptance. This explanation holds that the nature of terrorism has changed radically. Major terrorist attacks against the United States are no longer state-sponsored. Rather, it is claimed, terrorism is now carried out by individuals in "loose networks," amorphous, ill-structured groups, the existence of which may scarcely be known before they burst on the scene with a spectacular act of terrorism.[1] Usama bin Ladin and his organization, al-Qaeda, are but the most recent manifestation of this new terrorist phenomenon, said to have begun in 1993.

The assumption that terrorism is largely the work of isolated networks can make the determination of sponsorship much more difficult. When it was thought that terrorist states were behind most major acts of terror, authorities had a relatively short list of suspects to investigate. There was a realistic prospect of determining which state had been behind the attack and punishing it. Thus, except in wartime, the risk of truly major attacks on the United States had always been slim, for the prospect of ferocious retaliation was all too real for any nation to risk. But in an atmosphere in which almost anyone may be thought capable of carrying out clandestine attacks, the list of suspects is so long that the chances of swift reprisal are almost eliminated. As this "loose network" theory gained currency throughout the 1990s, it had a paralyzing effect on America's defensive stance.

This "loose network" concept of terrorism emerged out of two major bombing conspiracies in New York City in the first half of 1993. The first of those plots was the February 26, 1993, bombing of the World Trade Center. The mastermind of that plot, known as Ramzi Yousef, intended to bring both towers down, and though in practice his bombing fell short, the violence of his intentions was startling. In May 1994, in a stern address to the first four men to be convicted for that bombing, Judge Kevin Duffy reviewed the conspirators' aims at their sentencing hearing: to cause the North Tower to topple onto the South Tower amid a cloud of cyanide gas that would engulf those trapped in the first tower. "That's clearly what you intended," Duffy explained. "If that had happened, we would have been dealing with tens of thousands of deaths."

The Trade Center bombing was followed later that spring by a bombing conspiracy targeting the United Nations, New York's Federal Building, and two New York City tunnels. With the exposure of these two bombing conspiracies, the public—and the government—registered for the first time that loosely organized foreign conspiracies were targeting the United States.

The existence of such networks is now universally recognized. But it has overshadowed the possibility that rogue nations may be using those networks to achieve their own ends. Many important members of the 1993 World Trade Center bombing investigation, for example—particularly the New York FBI and its head, Jim Fox—believed that Saddam Hussein's Iraq was behind the Trade Center bombing. Fox arrived at the conclusion that it was a "false flag" operation in which the Muslim extremists who participated were left behind to be arrested and take the fall.

The second New York bombing conspiracy is similarly misunderstood. It was actually a sting operation that the FBI initiated to teach the Muslim extremists a lesson. The FBI used an informant who proposed making further bombs. A Sudanese émigré picked up the bait to make jihad and, as the plot progressed, the conspirators were eventually caught in the act of mixing what they thought was a bomb and arrested. Shaykh Omar and ten others were convicted for that plot. Once again, Washington did not pursue the issue of a sponsor state's involvement with the individual perpetrators. In fact, Sudanese intelligence had been involved in choosing the targets for the conspirators, and the distinct possibility exists that Sudan was fronting for Iraq.

Yet the involvement of states in those two bombing conspiracies, or at least their suspected involvement, received scant attention at the time. Instead, attention was focused on the individual perpetrators arrested for the bombing conspiracies and on the criminal proceedings that followed. Perhaps if we had given more consideration to the national security question of state involvement in such massive bombing plots on U.S. soil, subsequent attacks would not have occurred.

Instead, the 1993 bombing conspiracies were followed by another mega-terrorist plot against the United States in January 1995. Ramzi Yousef, who had succeeded in fleeing New

York the night of the Trade Center attack, reemerged two years later in the Philippines. He was plotting to bomb a dozen U.S. commercial aircraft with a liquid explosive that he could get past airport metal detectors. But while mixing chemicals, he accidentally started a fire. Forced to flee abruptly, he was arrested a month later.

After Yousef's second plot, America focused again on the criminal trial of the individual perpetrators and again paid little attention to the possible involvement of a state. Indeed, even though Yousef was eventually revealed *not* to be an Islamic militant, the theory of "loose networks" was only reinforced by his investigation and prosecution. As I will suggest, however, that theory is flawed—and fatally so. And its widespread acceptance in government and public circles only set the stage for the terrible events of September 11, 2001. In fact, that assault—the most lethal, cataclysmic terrorist attack in human history—must now be recognized as the inevitable escalation of the concerted campaign of terror that began in 1993.

An investigation into a major bombing conspiracy is a long, tedious affair. By its very nature, it does not produce decisive results quickly. It wasn't until several weeks after the 1993 Trade Center bombing that the FBI learned of the existence of the mysterious mastermind Ramzi Yousef. And they did so only because one of the conspirators, Mahmud Abu Halima, fled to his native Egypt, where he was arrested by Egyptian police and revealed Yousef's existence during a harsh interrogation.

The investigation into the September 11, 2001, assault is particularly difficult because there were four separate attacks. Moreover, the individuals who carried them out are all dead; no information can be gained from them. It may be a long time, probably years, before the inquiry into those terrible events leads to a clear, indisputable conclusion. It may even be that reaching an indisputable conclusion will ultimately prove impossible. But we must not wait to act. We must use what we know from the past. We must scrutinize what we know not only about the current attacks, but also about those that preceded them, in order to understand what may lie behind them.

Fortunately, we have on record a good deal of information from the first World Trade Center bombing. That evidence was prepared and made public by two men for whom I came to have the greatest regard. One was Jim Fox, the head of New York FBI, who led the investigation in New York and who reached the conclusion that Iraq was behind the Trade Center bombing.[2] In the course of my research for this book, we became friends. Fox passed away suddenly in the spring of 1997, when he was just fifty-nine. His untimely death was a great loss. The other man responsible for preparing the evidence for the first Trade Center bombing trial was Gil Childers, the lead prosecutor in that trial. Childers put into the public record all the information necessary to demonstrate to the jury that the four defendants on trial and the two indicted fugitives, one of whom was Ramzi Yousef, had done what the government claimed—participated in a conspiracy to topple New York's tallest tower onto its twin. And Childers did so with enormous care and detail.

Their thorough work allows us to raise questions about what really happened in 1993, and what is still happening today. Reviewing it firsthand allows us to reevaluate the unanswered questions surrounding the first World Trade Center bombing, to see what Washington missed then, and what it may still be missing.

Who is Ramzi Yousef, and who was behind his acts? Who funded him? Who trained him? These questions were never investigated properly because of a peculiar division that exists in America between the Justice Department and the national security agencies—the State Department, Pentagon, and CIA. There is an organizational firewall between them that prevents one from "interfering" in the work of the other. Above all, the national security bureaucracies are prohibited from interfering in the work of the Justice Department, for that could be considered obstruction of justice.

But what happens when a terrorist act falls within both spheres? Terrorism when committed by individuals is a crime. When supported by states, however, terrorism is also war. What happens when a state commits an act of terrorism in America? Is that a criminal matter or a national security issue?

The two questions are often interrelated, but the American government treats them as separate issues. The Justice Department can try individual perpetrators or the Defense Department can punish a state sponsor. But both cannot be done at the same time under current bureaucratic procedures. The chief business of the Justice Department is to conduct trials—to prosecute and convict individuals. Once an arrest is made, the Justice Department declares the matter *sub judice* and denies information to the national security bureaucracies. But this creates an organizational firewall, because the national security bureaucracies are responsible for determining whether any act of terrorism had state sponsorship. Their job is much more difficult without the relevant information from the Justice Department. The Justice Department, however, places a higher priority on the prosecution and conviction of individuals—with all the rights to a fair trial guaranteed by the American legal system—than on the country's national security interest in determining who might have been behind the terrorists. Indeed, the question of state sponsorship may be relegated to such a distant second place that it may never be addressed at all.

I first understood this problem from the federal prosecutors in the New York bombing conspiracies. The unusual meeting was arranged by the New York District Attorney's office, and it included Andrew McCarthy, the lead prosecutor in the conspiracy trial of Shaykh Omar et al., and Gil Childers, lead prosecutor in the first World Trade Center bombing trial. When I explained to them that Iraq was behind the World Trade Center bombing and that Ramzi Yousef was almost certainly an Iraqi intelligence agent, they replied, "You may be right. But we prosecute individuals. We don't do state sponsorship." [3]

"Then who 'did' state sponsorship?" I asked the New York prosecutors. "Washington," they replied. "Who in Washington?" I asked. They said nothing. The room was silent.

The answer, it turns out, is that Washington did not, in fact, address the question of state sponsorship (or it did not do so properly, at least). The Justice Department was the only bureaucracy that had both the evidence, produced by the FBI in the course of its investigation, and the intelligence, produced by the CIA and other national security bureaucracies in the course of their inquiries. The FBI's National Security Division was formally responsible for investigating the question of state sponsorship.

But the FBI is primarily a domestic law enforcement agency. Following the bombing of Pan Am Flight 103, which crashed over Lockerbie, Scotland, in December 1988, the Scottish police were the lead investigative agency. They made their findings available to both the CIA and the FBI. Thus, in the case of Pan Am Flight 103, there were two investigations: an intelligence investigation, led by the CIA, focused on the question of state sponsorship; and a criminal investigation, led by the FBI, focused on individual perpetrators. *But there was never a properly conducted intelligence investigation of the World Trade Center bombing.*

The national security bureaucracies had only foreign intelligence to decide the question of state sponsorship. But many cases of state-sponsored terrorism cannot be resolved with intelligence alone. Intelligence did not link the bombing of Pan Am Flight 103 to Libya. It was *evidence*—a microchip, part of the bomb's timing device—that revealed the connection to other bombs built by Libya.

The FBI, then, was the only bureaucracy that had all the information relevant to the World Trade Center bombing, both the evidence and the intelligence. But even if the FBI *did* make a serious effort to examine the evidence for state sponsorship—and it is not clear that it did—the FBI alone is probably not qualified to carry out such an investigation.

"They're headhunters," one official in the Pentagon's counterterrorism office remarked[4]—that is, the FBI is oriented toward arresting individuals rather than pursuing state sponsorship. Moreover, the FBI lacks Middle East expertise. One State Department expert said of the FBI's office of radical Islamic fundamentalism, "It's a joke."[5]

The investigations into the Pan Am Flight 103 bombing were among the largest criminal and intelligence investigations in U.S. history. It took more than a year and a half to find the key evidence—a microchip, half the size of a man's thumbnail—amid the plane's wreckage scattered over many miles of Scottish countryside. No such comparable effort was made to determine whether the World Trade Center bombing had state sponsorship. Instead, America held a trial and otherwise left the issue to the Justice Department, which was oriented to the prosecution and conviction of individuals.

ABC and *Newsweek* conducted a three-month investigation into the World Trade Center bombing in 1994, for which I was a consultant. This book is based on that initial investigation, together with work I have done since and documents issued by the U.S. Attorney's Office of the Southern District of New York. But, above all, this book is based on the evidence from the first World Trade Center bombing trial.

Although the national security bureaucracies were never given that evidence, evidence becomes public during a trial. The World Trade Center evidence consists of thousands of pages of raw data, such as telephone records, passports, and airplane tickets. In itself, the evidence reveals nothing about state sponsorship. But, when carefully analyzed, it provides a broader picture.

The great weight of the evidence, I believe, points unmistakably to the conclusion that the World Trade Center attacks were state-sponsored. By treating the 1993 World Trade Center bombing and subsequent acts of terrorism strictly as criminal issues, Washington left America exposed and vulnerable to more terrorism from Saddam Hussein or any other party that might choose to adopt the audacious winning strategy. By sacrificing minor collaborators to be tried by the Justice Department, states promoting terrorism can conceal their role and operate with impunity.

A proper understanding of the first World Trade Center bombing further suggests that several other major acts of terrorism directed against America probably also had state sponsorship. That understanding places the events of September 11, 2001, in a dramatically new perspective. The question of Iraq's involvement, and of Saddam Hussein's unfinished war against America, has special urgency today.

2

The Assassination of Meir Kahane

The World Trade Center bombing had its immediate origins in the fulminations of a radical Islamic Egyptian immigrant, El Sayyid Nosair. He had been convicted in December 1991 on assault and gun charges related to the 1990 assassination of Meir Kahane, the right-wing American-Israeli leader of the Jewish Defense League.

El Sayyid Nosair

By any standards, Nosair was a bit nutty. Born in 1955 near the city of Port Said, which commands entry into the Suez Canal from the Mediterranean Sea, Nosair was the oldest son among the five children of an electrical power plant foreman.

Like those of so many Middle Easterners, Nosair's life was soon caught up in the region's endemic violence. After Egypt lost the Sinai Peninsula to Israel in the 1967 war, Egyptian president Gamal Abdel Nasser launched the "War of Attrition." Nasser was unwilling to make peace with Israel, and the War of Attrition was an attempt to force Israel to withdraw unconditionally, by inflicting unacceptable casualties on front-line troops positioned on the eastern bank of the Suez Canal. Nosair's family, along with hundreds of thousands of others living in the canal zone, were relocated to Cairo. He grew up a country boy in the big city.

Nosair graduated in 1978 from Cairo's College of Applied Arts with a degree in industrial design engineering. But he soon suffered a personal tragedy. His father had returned to Port Said to work there, leaving the family in Cairo. Suddenly and unexpectedly, Nosair's mother died, in his arms.

Shortly thereafter, Nosair left for America, arriving in 1981 to take up residence in Pittsburgh, where a close family friend lived. He found employment as an apprentice gem-setter and repairman at a family-run jewelry store.

Nosair was not then a religious extremist. Rather, as his first employer explained, "He loved everything about this country. . . . He'd go to the zoo, museums, nightclubs. He really got into fast food. And he was very intrigued by young women. . . . He liked to tell jokes and did a funny yukyuk laugh. It was hilarious. You got the biggest kick out of him." But, as she also added, "He had a very high opinion of himself. His goal was to get rich and return to Egypt."[1]

Within a year of arriving in America, Nosair quarreled with his host and turned toward Islamic extremism. Through a matchmaker at Pittsburgh's Dar al-Salaam [House of Peace] mosque, he met a recent divorcee, Caren Ann Mills, who had converted from Catholicism to Islam as she sought to cope with the emotional strain of her divorce. Nosair married Mills in June 1982, a few months after meeting her. He had overstayed his visa and candidly admitted that he was looking for an American woman to marry so he could obtain a green card.

At the same time, Nosair's work was becoming increasingly unsatisfactory. He repeatedly broke costly stones. That, along with his growing religious zealotry, led to his being fired early in 1983, after which he supported his family with unemployment compensation, welfare, and odd jobs.

Despite his zealotry, however, Nosair's life among the extremists was not smooth. In 1985 he was accused of rape by a young woman, newly converted to Islam, staying in his house. The woman filed a formal complaint with the police. Two weeks later another woman filed an assault complaint against him.

Members of the Dar al-Salaam mosque, to which the two women and Nosair all belonged, persuaded the women to drop their complaints. Nosair left Pittsburgh, moving to the New York area, where an older cousin, Ibrahim El Gabrowny, lived.

In New York, Nosair found employment as an electrician's aide with a private company that did work at an electrical sub-station owned by the Port Authority of New York and New Jersey. In September 1986 he was seriously injured in an electrical accident. Nosair underwent repeated skin grafts and was bedridden for six months, having suffered severe burns, neurological damage, and a spinal sprain. He was also left impotent. He began seeing a psychiatrist and taking medication, including the controversial drug Prozac, to treat his depression. In 1988 Nosair sued the Port Authority in a civil suit that remained pending at the time of Kahane's murder.

Nosair worshiped at the Masjid al-Salaam, the inaptly named Mosque of Peace, a dingy assortment of rooms over a Chinese takeout establishment in Jersey City. One of the smaller, poorer, and most radical mosques in the New York area, the Masjid al-Salaam was founded by a Palestinian from Gaza, Sultan El Gawli, who had immigrated to the United States in 1968. Much later Shakyh Omar Abdul Rahman was to preach there, after settling in America.

El Gawli was arrested in December 1985, in a U.S. customs sting operation. At a November 12, 1985, meeting at a travel agency that El Gawli owned, an Egyptian who worked as a U.S. informant met with the PLO representative to Saudi Arabia, Said Hassan. Hassan asked the informant to supply him with 150 pounds of the explosive C-4, along with blasting caps and a remote detonator. The PLO was to use the material in Israel over the Christmas season, and El Gawli was to receive $25,000 for his role in arranging the shipment. El Gawli was convicted of conspiracy to export explosives in July 1986 and sent to federal prison. He was released in October 1990, a month before Kahane's assassination.

The Murder of Meir Kahane

Meir Kahane was the son of a New York rabbi with right-wing political convictions, a close friend of the revisionist Zionist leader Ze'ev Jabotinsky. Kahane was first arrested in 1947, at the age of fifteen. He had participated in a protest against the

British occupation of Palestine, throwing rocks at the limousine of the British foreign secretary as he visited New York.

In 1968 Meir Kahane, then an unknown rabbi, together with a small group of radical Jews founded the paramilitary Jewish Defense League (JDL). Arrested many times in New York on charges of participating in disorderly protests, in July 1971 Kahane was given a five-year suspended sentence and fined $5,000 after being convicted for conspiring to manufacture explosives. Two months later he fled to Israel. In 1975 he was returned to the United States to serve a year in federal prison for parole violations.

In 1984 Kahane founded Kach, Hebrew for "Thus," a political party that advocated the expulsion of all Arabs from the occupied territories and Israel itself. After two previous failed attempts to win a seat in the Israeli Knesset, Kahane finally succeeded in his bid in 1984. As a parliamentarian, he drafted legislation reminiscent of the Nuremberg Laws, making it a crime, punishable by two years in prison, for a Jew to have sex with an Arab. The Knesset soon passed a law banning parties with racist platforms. An Israeli court found him ineligible to run in the upcoming 1988 elections and Kahane's brief parliamentary career ended.

On the evening of November 5, 1990, Kahane addressed some sixty members of ZEERO, the Zionist Emergency Evacuation Rescue Operation, which was holding its founding conference at a hotel in midtown Manhattan. Kahane was assassinated while answering questions after his speech. El Sayyid Nosair fled the murder scene, scuffling with, shooting, and seriously wounding a seventy-three-year-old man who tried to stop him.

Nosair then commandeered a taxicab at gunpoint, causing its Hispanic driver, as the man later testified, to "get so scared I make wet in my pants."[2] But the cab became stopped in traffic after traveling only seventy-five feet, and Nosair panicked, leaping out in front of a post office. As he started to flee, he was confronted by a federal postal inspector. Nosair shot him in the chest. The inspector, who was wearing a bulletproof vest, was only lightly wounded. He returned fire, hitting Nosair in the jugular vein and felling him.

Nosair was rushed to Bellevue Hospital in critical condition. In an unusual bedside arraignment, as he lay in intensive care, attached to tubes and breathing through an oxygen mask, Nosair was charged, November 7, with second-degree murder, attempted murder, assault, and criminal possession of weapons.

The Trial of El Sayyid Nosair

Nosair's trial began a year later, in November 1991. Local Muslim extremists, many from Nosair's mosque, the Masjid al-Salaam, turned out to support him, while Kahane supporters turned out against him. Prominent among the leaders of the anti-Nosair group was Dov Hykind, a New York assemblyman who represented an Orthodox Jewish district in Brooklyn.

William Kunstler, seventy-five years old, headed Nosair's legal team. Kunstler was known for his defense of radical, unpopular, and even outrageous defendants. He was also an outspoken critic of Israel and had been given the honorary title "Moses Muhammad" by a Brooklyn Islamic cleric several years before.

Michael Warren was another of Nosair's lawyers. A black Muslim with close ties to Louis Farrakhan's Nation of Islam, Warren had been involved in the racially charged Tawana Brawley hoax several years before. According to investigative journalist Robert Friedman, Nosair's "virulent anti-Semitism [was] matched by that of Warren" himself.[3]

The prosecution had such an apparently solid case against Nosair that in the beginning his legal team wanted to use a temporary insanity defense. As Kunstler himself explained, "When I first took on Nosair, I believed he was guilty."[4]

But from the very beginning, even as he lay in intensive care after the shooting, breathing through a tube in his throat and unable to speak, Nosair insisted that he was innocent. Someone else had committed the crimes. During his arraignment, Nosair gave police a 250-word statement in which he asserted that the only reason he fled after Kahane's murder was that as the sole Arab at the ZEERO meeting he feared he would be blamed for the crime. A man wearing a black yarmulke, Nosair claimed, was the person who actually killed Kahane and shot the postal inspector:

> When the man with the yarmulke was right behind me, I saw a police officer was hiding in a front door of a building. So I run towards him to ask him to help me—the other man was right behind me with a silver gun when

I reached to the policeman. The man behind me shot many shots so the policeman started to shoot towards us. When I got shot the man with the yarmulke stood beside me. When I was laying on the ground, he put the gun beside me.[5]

Nosair's lawyers immediately disavowed Nosair's statement, insisting that he was too heavily sedated to have known what he was saying.

But eventually the defense team went with Nosair's story, although it was entirely invented, dropping plans for an insanity defense in July 1991. They made the best of what they had, sometimes using quite unscrupulous tactics in the process. Michael Warren repeatedly charged that the judge, Alvin Schlesinger, who was Jewish, was "bought off" by Zionist interests and was "colluding with Zionist interests."[6] Nosair's supporters demonstrated outside the courtroom with placards proclaiming, "We don't want a Jew judge."[7] They either did not realize or did not care that their own favorite lawyer, William Kunstler, was also Jewish, as they vented prejudice that, if expressed against them, would have prompted their indignant outrage.

Although the defense had a weak position, it did not entirely lack points to exploit. There were sixteen witnesses to Kahane's assassination, but they were all watching Kahane at the time, not Nosair. None actually saw Nosair pull the trigger. Nosair's lawyers emphasized discrepancies in their testimony while hinting that Kahane was killed by one of his own supporters after he discovered that money had been skimmed off a JDL bank account in New York.

That claim, like Nosair's story of the man with the black yarmulke, was a complete fabrication. As the *National Law Journal* noted after the trial, the defense "never produced any evidence to support this theory," and during the trial Justice Schlesinger was obliged to bar the defense from using "pure speculation" as it questioned witnesses about a killer other than Nosair.[8]

By the trial's end, the prosecution's case looked very strong. One witness testified that he saw Nosair standing near Kahane with a gun in his hand right after the shooting; another said that he saw Nosair flee the room carrying a gun. The postal inspector testified, "This man right here shot me," graphically describing the expression on Nosair's face then.[9] A police ballistics expert explained that the bullets that killed Kahane and wounded the two other victims could only have come from the gun found at Nosair's side. Kunstler was pessimistic as the case went to the jury on December 19.

But after four days of sometimes acrimonious deliberations—a female juror complained that a male juror had called her a "jackass"[10]—the jury returned with a stunning verdict. It acquitted Nosair of charges of murder and attempted murder and convicted him only of assault and weapons charges. The jurors were not convinced that Nosair fired the shot that killed Kahane, and they believed that in shooting the postal inspector Nosair had not meant to kill him.

In essence, the jury, with very little reason, rejected the authority of the government ballistics expert, along with the testimony of those who saw Kahane's murder and Nosair's flight. At the January 29, 1992, sentencing, Justice Schlesinger pronounced the verdicts "against the overwhelming weight of evidence" and "devoid of common sense and logic."[11] He gave Nosair the maximum sentence possible, seven and one-third to twenty-two years. Nosair was sent to Attica penitentiary. He would be eligible for parole in six years.

The racial composition of the jury played a key role in the unexpected verdict, or so many people thought, including Kunstler. After the verdict, Kunstler boasted that the selection of jurors was the defense team's critical move. We had worked, he said, for "third-world people . . . people who were not yuppies or establishment types."[12] Indeed, in the middle of jury selection, the prosecution had charged that the defense was rejecting candidates on the basis of race, and it moved for a mistrial.

The defense had used ten of its allotted twenty peremptory challenges to exclude whites. Of the first six jurors chosen, five were black and one Hispanic. Justice Schlesinger ruled against the call for a mistrial but ordered that both sides would have to show that race was not a factor in choosing future jurors. Five whites and one black were subsequently added to the jury. But it was too late. As one legal expert commented after the verdicts, "Apparently the prosecutors didn't do too good a job picking a jury. If all cases were held up to these standards, 90 percent of the people accused of crimes would go free just because there's not an eyewitness. People don't commit rape in a Macy's window."[13]

Links to Terrorist Organizations?

Already by November 7, 1990, two days after the shooting, New York authorities insisted that Nosair had acted alone. As Chief of Detectives Joseph Borelli explained, there was no evidence linking Nosair to any anti-Israeli organization or terrorist group. "I'm strongly convinced that he acted alone. . . . He didn't seem to be part of a conspiracy or any terrorist organization."[14]

Kahane was murdered in the midst of the Persian Gulf crisis, as America prepared to go to war with Iraq. Tensions were high. The day after Kahane's assassination, as his supporters cried for revenge, two elderly Palestinians were shot and killed on the West Bank, a sixty-five-year-old man riding a donkey to his olive fields and a sixty-year-old woman standing in a nearby doorway. The month before, nineteen Palestinians had been killed by Israeli troops in a riot on the Temple Mount, as they threw stones on Jewish worshipers praying at the Wailing Wall below.

It is understandable, then, why authorities may have wanted to calm tempers by insisting that Nosair was a lone gunman and playing down the political aspects of Kahane's assassination. But only two days after the murder they could not possibly have conducted the investigation that would have allowed them to conclude that Nosair had acted alone. Moreover, as would be revealed some five years later, during the trial of Shaykh Omar et al., authorities did have reason to believe that Nosair was part of a larger group. He and his comrades had been under surveillance at least since 1989, when the FBI photographed Nosair, along with three others who would later be convicted of the World Trade Center bombing and a fourth who would stand trial along with Shaykh Omar, practicing together at a Long Island gun range, firing AK-47 rifles and 9-millimeter semiautomatic pistols.

Given what authorities knew about Nosair, it would have been reasonable to assume that others were probably involved in Kahane's assassination, even if, perhaps because of tensions related to the gulf crisis, they did not want to say so then. But when Nosair's trial began a year later, after sufficient time had passed to conduct a thorough investigation and after the political atmosphere had returned to normal, authorities still maintained the same position. They continued to claim that "there was no evidence linking the defendant to any Mideast terror organization or suggesting that it was part of any larger conspiracy."[15]

But the position that Nosair had acted alone was not to be the government's last word. Indeed, already in December 1991, in the middle of Nosair's trial, one of the prosecutors reported that "a note found among Nosair's possessions 'indicates that someone else may have been involved.' "[16] Still the issue was not pursued.

Only two years later, six months after the World Trade Center bombing, would authorities finally conclude that others had been involved in Kahane's murder. Mahmud Abu Halima—also an Egyptian fundamentalist, and Nosair's good friend—had been the intended getaway driver. But Abu Halima had waited in the wrong spot, causing Nosair to panic and commandeer a taxi.

Moreover, Kahane's murder, federal authorities would charge in 1993, was part of something much bigger. It was the first criminal act in a seditious conspiracy led by Shaykh Omar Abdul Rahman against the government of the United States. Thus, Kahane's murder was not adequately explained during Nosair's trial in 1991. But, even now, has it been adequately explained?

The Abu Nidal Organization, which also calls itself the Fatah Revolutionary Council (FRC), was established in Baghdad by a dissident Palestinian, Sabri al Banna. Born in Jaffa in 1937, al Banna was the son of the eighth wife of a wealthy Palestinian landowner, who died when al Banna was eight years old. As a result of the 1948 Arab-Israeli war and the establishment of the State of Israel, the family lost all its property. Al Banna was raised by his mother and older brothers in Nablus, in the West Bank, then under Jordanian control, in impoverished circumstances.

An indifferent student, al Banna started but never finished college. Instead, he went to work as a laborer in Saudi Arabia in the early 1960s. He was affiliated with the radical Arab nationalist Ba'th Party, which was banned in Saudi Arabia. After 1967 al Banna, whose nom de guerre was Abu Nidal, joined Yasir Arafat's Fatah and was made the PLO's "ambassador" to Baghdad, where the Ba'th seized power in 1968.

Abu Nidal opposed the peace talks that the Arab states began to conduct with Israel in the wake of the October 1973 war. So did Iraq, and Baghdad helped establish his organization. Iraqi intelligence, itself trained by the Soviets, trained Abu Nidal's men.

The FRC quickly became an instrument of Iraqi policy. In its first years, the FRC's principal targets were Baghdad's Arab rivals. Only in the late 1970s did Abu Nidal begin to attack Jewish and Israeli targets. Like his mentors, Abu Nidal is unusually cruel and ruthless. The FRC was responsible for some of the most bloody and indiscriminate terrorist acts of the 1970s and 1980s, including the December 1985 assaults at the El Al ticket counters of the Vienna and Rome airports, which killed eighteen people and wounded forty others.

In 1979, as Abu Nidal began to attack Jewish and Israeli targets, the FRC started to shift the focus of its activity from the Middle East to Western Europe, where it began to establish secret cells. The FRC controlled money for scholarships for Palestinian students to attend schools in Europe. In return, the students agreed to be "sleepers" for the organization. That is, they would work for the FRC. But they would not act immediately. They would establish themselves abroad. And when their services were needed, perhaps years later, they would support whatever operation the FRC wanted to undertake.[17]

One such cell was responsible for the attempted assassination of Israel's ambassador to London on June 3, 1982. It left the ambassador, Shlomo Argov, permanently paralyzed.

Marwan al Banna, a second cousin of Abu Nidal, had failed to gain admission to a college in Syria where he wanted to study. In November 1979 he was sent to London on an FRC fellowship. There he established an FRC cell, gathered intelligence, and otherwise went about his life as a student. The idea was that when the time came, Marwan al Banna would be in a position to arrange the logistical aspects of the operation, handle necessary communications, and take care of finances. Al Banna was furnished with codes and passwords by which he would be able to identify couriers and contacts. He was given two numbers for use in emergencies: a telephone number and post office box in Baghdad. After spending more than two years in England, al Banna was finally called upon to act. His group was tasked with assisting a colonel in the Iraqi intelligence service in assassinating Shlomo Argov.

The would-be assassins were arrested, tried, and convicted in March 1983 in a British court. The proceedings provided rare insight into the structure and operations of a Middle Eastern terrorist cell. It will be helpful to bear in mind the structure of that cell, when we later consider the World Trade Center bombing.

On December 5, 1990, Justice Schlesinger set Nosair's bail at $300,000. Two days later, on December 7, the FRC in Beirut announced that to honor Nosair and commemorate the anniversary of the Palestinian uprising against Israel, which had begun on that day three years earlier, it had "taken a decision to bear the defense expenses of the heroic Arab struggler El Sayyid Nosair."[18] The FRC would provide the $300,000 for bail. Upon further consideration, and after learning that Nosair had passports in two names, Justice Schlesinger revoked bail.

Indeed, Nosair had a variety of documents. He had a New York license plate inside his car, with New Jersey plates on the outside. He had three New Jersey driver's licenses, each at a different address. He habitually used different variations of his name: Noseir, Nosir, and Nasser. On his application for naturalization as a U.S. citizen, he identified himself as El Sayyd Abdulaziz El Sayyd. On city employment records, he was listed as El Sayyd Nosair.

Was there some link between Nosair and the FRC, or at least one of its members? In April 1986 a Palestinian with American citizenship, along with two other Palestinians, both cousins, attacked an Israeli bus on the West Bank. The driver was killed and a passenger seriously wounded. The two cousins were arrested, but Mahmud Atta fled.

One of the two men arrested for the bus attack broke and confessed to Israeli authorities that Atta was a member of the FRC. U.S. investigators subsequently identified Atta as chief of Abu Nidal's operations in the United States and Latin America.

Israeli and U.S. authorities, working together, monitored a

meeting in April 1987 in Mexico City between Atta and four Palestinian-Americans from several Midwestern states. Atta instructed them on how to raise money in the United States to be sent to Abu Nidal overseas and gave them a variety of other tasks. After the meeting, Atta returned to Venezuela, the others to the United States.

Atta was seized in Venezuela and sent to the United States, to New York. He was to be extradited to Israel, and he appealed the extradition order. On August 31, 1990, a U.S. Court of Appeals rejected his final bid. The FRC repeatedly warned of retaliation if Atta were extradited. Indeed, five days before his October 30 extradition, the U.S. State Department issued a statement warning of a possible Abu Nidal attack in the Middle East.[19]

Two days after Atta's extradition, the FRC issued a statement denouncing the move, claiming that Atta was a political prisoner, while asserting that international conventions prohibited the extradition of political prisoners. The FRC again threatened retaliation, warning that it would hold the United States responsible for Atta's safety.[20]

Four days later, Nosair shot Kahane. Among Nosair's possessions was found a list of five Jewish names scrawled on a page from his address book. They included a congressman, a newspaper columnist, two federal judges, and a former assistant U.S. attorney. The two judges and attorney had all been involved in Atta's extradition. Investigators concluded that it was a hit list.

At the time, the Midwestern Abu Nidal cell to which Atta had given instructions still existed, while others in America had recently been rounded up. Investigators had allowed Atta's organization to continue operating because of the information they got from monitoring it. However, in 1989 one of the members of the cell and his wife were arrested for murdering their sixteen-year-old daughter. They feared that family tensions might lead her to reveal their activities. FBI listening devices picked up the girl's screams as she was killed in her house by her parents.

As her father came at her with a nine-inch butcher knife, she cried, "Mother, please help me." "What do you mean, what help?" replied the 224-pound woman, as she wrestled her daughter to the floor, holding her by the hair, while her husband repeatedly stabbed her. Near the end, panting from his efforts, the father whispered, "Die! Die quickly. . . . Die my daughter die."[21]

The three other members of the Midwestern cell were arrested in April 1993. After the World Trade Center bombing, it was judged too dangerous to leave such people free.

Atta was extradited to Israel on October 30, 1990. Two days later the FRC threatened retaliation, and four days after that Nosair shot Kahane. The FRC offered to post Nosair's bail, and the names of three people involved in Atta's extradition were found on a hit list in Nosair's apartment. Was Kahane's assassination motivated, at least in part, by a desire to take revenge for the extradition of a member of Abu Nidal's organization?

Although Mahmud Atta was not an Islamic extremist, the New York extremists would have sympathized with him. When Nosair finally acted on the violent emotions he had long nurtured, could Atta's extradition have provided the occasion for the expression of his hatred? There would seem to be sufficient circumstantial evidence suggesting a link between Atta's extradition to Israel and Nosair's assassination of Kahane not to dismiss the idea out of hand. But that is what authorities did.

"There Is No Evidence"

Often, when American officials say they have no evidence, they mean that they do not have proof; they do not have a smoking gun. But that is a misuse of the word *evidence*. There is a difference between evidence and proof. Webster's dictionary defines evidence as "something that indicates," using the example, "your reaction was evidence of innocence." Proof is something different: "evidence establishing the validity of a given assertion" or "conclusive demonstration." Evidence often has to be developed and aggressively pursued until it becomes proof.

The repeated claim that authorities had "no evidence" linking Nosair to any terrorist organization was not true. Nosair had clear links to Muslim extremists, and there was evidence— "something that indicates"—of a link between Nosair's murder of Kahane and Atta's extradition to Israel. But only a few people at the time, mostly Kahane's supporters, publicly challenged the idea that Nosair was a lone gunman.[22]

And yet, perhaps more people should have challenged it. The World Trade Center bombing conspiracy soon emerged from the assassination of Meir Kahane. Had authorities not insisted that Nosair acted alone, had they acknowledged that he had at least one other accomplice and, in any event, lived in a violent environment, perhaps they would have dealt with him differently. Perhaps they would have placed him under tighter security in prison. Perhaps they would have been more careful in supervising the informant who infiltrated the group around him. Perhaps many things would have been done differently.

But how, precisely, did the assassination of Meir Kahane by a fanatic, somewhat unbalanced Muslim extremist lead to the World Trade Center bombing?

3

The Origins of the World Trade Center Bombing

In the spring of 1992, El Sayyid Nosair brooded in Attica Penitentiary. His thoughts focused on three things:

1. He wanted revenge. He wanted to assassinate Justice Schlesinger, who had denounced the "innocent" verdicts on the most serious charges he had faced, and then given him the maximum sentence on the "guilty" verdicts that the prosecution had managed to secure. Nosair also wanted to kill New York assemblyman Dov Hykind, the most prominent figure among those who turned out at the courtroom to denounce him.

2. He wanted to escape from prison.

3. And he wanted his friends to do something. He resented the fact that they were free, doing little, as he saw it, to further the cause about which he felt so passionately, while he had gone to jail for Islam.

Those who had supported Nosair during his politically charged trial continued to visit him regularly in Attica. They included his cousin, Ibrahim El Gabrowny, and a third Egyptian, Emad Salem.

Emad Salem: A Tale of Two Masters

At the time, Salem was an FBI informant. He was also a long-time Egyptian intelligence agent with "resilient ties" to his former employer.[1] Indeed, investigators nicknamed him "the Colonel," in allusion to those ties.

The Egyptian government had originally sent Salem to America because it was concerned about the activities of its Egyptian opponents in the United States, according to Vincent Cannistraro, former director of the CIA's office of counterterrorism operations.[2] Salem had arrived in the United States in September 1987. He soon met an American woman, Barbara Rogers, at a Tae Kwan Do studio in Manhattan, where they both worked out. She was a thirty-three-year-old marketing employee for Avon products. Although Salem already had a wife and children in Egypt, he married Rogers six weeks later, on November 8, 1987.

In New York, Salem held an irregular series of jobs. First he worked as a security guard, then as a taxicab driver, and then as a security guard again, when he discovered how much he resented the rudeness with which his passengers sometimes treated him. He was chronically short of money. He wanted a regular salary from the FBI, but it would only pay him by the piece for his information.

Salem's continuing ties to Egyptian authorities manifested themselves in several ways. According to Rogers, even before coming to America her husband had agreed to report to Cairo any contact he might have with five members of the Egyptian army who had failed to return home after training here. Salem made frequent trips to Egypt, three times in 1988 and twice in 1989. Rogers accompanied him on one trip in March 1990. They were treated "like royalty," met by security officials at the airport and housed in special military quarters.[3]

Because of his marriage to an American, Salem was entitled to accelerated citizenship, and he became a U.S. citizen in August 1991. But, at the same time, his marriage to Rogers was not going well. The two were soon separated, and Salem began dating a German woman, Karen Goodlive. A convert to Buddhism, Goodlive helped market "You're Gonna Lov' It Designs," a costume jewelry collection.

Salem and Rogers were divorced the next year. It was a bitter proceeding. In a February 1992 affidavit, Rogers charged that

Salem "threatened me, abused me, cheated on me and generally treated me in a manner that endangers my physical and mental well-being." She also claimed that Salem had "threatened to kill my cat and to cause me further harm" if she refused to approve an uncontested divorce. In his reply affidavit, Salem denied the "absurd statements."[4]

In November 1991, Salem had an accident at the Woodward Hotel in Manhattan, where he worked as a security guard. He fell off a ladder, landing on his head. According to his supervisor, Salem started acting belligerently. He began pressing co-workers to buy the jewelry his girlfriend sold, threatening them with his gun. He obtained a credit card in another guard's name and used it, without the man's knowledge. Salem was soon fired from his job. Four months later, he got even. In July 1992, he returned with a squad of immigration agents to arrest illegals working there.

In November 1991 Salem, on behalf of the FBI, and while still working with Egyptian intelligence, began to infiltrate the Muslim extremists who supported Nosair during his trial, focusing on the Egyptians among them. Being Egyptian himself, it was natural that Salem should do so, while the Egyptian government was most interested in those people. Under assault from Muslim extremists at home, including the followers of Shaykh Omar Abdul Rahman, the Egyptian government took a far more alarmist view than U.S. authorities did of Shaykh Omar et al. It believed that the Americans did not understand the real threat that he posed.

Cairo was disturbed by the freedom that Shaykh Omar enjoyed in America to communicate with his followers in Egypt, raise money, and otherwise rally support. But even more, Cairo suspected that in giving Shaykh Omar a haven, the United States was hedging its bets against a possible Muslim extremist takeover in Egypt. The U.S. embassy in Egypt had been holding meetings with members of Shaykh Omar's group since 1991. Indeed, their last meeting was held just six weeks before the World Trade Center blast.

Almost as soon as Salem became involved with the Muslim extremists in November 1991, Shaykh Omar solicited him to assassinate Egyptian president Husni Mubarak. Salem reported that to Egyptian authorities and otherwise gave them information about those around the shaykh.

From the time Salem first penetrated their ranks, the extremists talked about making bombs and springing Nosair from prison, but their plans did not begin to coalesce until some months after Nosair was convicted and sentenced, in May and June of 1992. In May 1992, at El Gabrowny's direction, Salem visited Nosair at Attica. Nosair pressed him to arrange for the assassination of Dov Hykind and Justice Schlesinger and discussed plans for his escape and for making bombs.[5]

El Gabrowny hesitated. He advised waiting for the results of Nosair's legal appeal. But El Gabrowny's resistance gradually eroded, and over the next two months the idea of assassinating individuals was dropped and a bombing plot emerged as the central feature of the conspiracy. El Gabrowny was to direct the operation, and Salem was to provide the technical expertise.

After El Gabrowny went to see Nosair at Attica on June 6, Salem went to Attica a week later, on June 14. Nosair pressed Salem to help with a bombing plot that he had conceived. He wanted to build twelve pipe bombs, some to be placed where Assemblyman Hykind and Justice Schlesinger would be among the victims, others to be used to attack random Jewish targets.[6]

Two days later, on June 16, Salem met with El Gabrowny in Brooklyn to continue their discussions about building the bombs.[7] Salem even started to buy parts for them.

Nosair's plot—the twelve pipe bombs—certainly would have implicated a lot of his friends. They too would be doing something for Islam. But it is not clear how the plot would have helped spring Nosair from jail, which Nosair also very much wanted. But then Nosair was more violent in his impulses than realistic about what he could achieve. A year later, in the spring of 1993, he would propose that another group of friends kidnap Henry Kissinger and Richard Nixon and hold them hostage for his release.

In any event, the conspirators looked to Salem, the FBI informant, to make Nosair's pipe bombs. Yet at that point—just as the plot had begun to take shape—a dispute emerged between the FBI and Salem over how to proceed.

Salem had been involved in an intelligence investigation. The FBI was gathering information on the Muslim extremists in the New York area. But once Salem proposed to his FBI handlers that he and the extremists actually build bombs, it became a criminal matter. Then, Salem would have to follow procedures for a criminal investigation. That meant that he would have to wear a body recorder to gather evidence and serve as a witness in any trial, if necessary. Salem refused.[8]

The FBI did not really trust Salem. He was, after all, a less than unimpeachable character, and the FBI suspected his ties to Egyptian intelligence. Salem took three lie detector tests in the spring and summer of 1992. The first was inconclusive, but the other two suggested that he was not truthful with authorities.[9]

Finally, as one of his FBI handlers later explained to Salem, his view of the seriousness of the plot was "pooh-poohed" by superiors.[10] Indeed, that view was not without merit. Without Salem, or a figure like him, the conspirators did not know how to build a bomb, not even small pipe bombs. The FBI dropped Salem in early July 1992. Salem's contacts with the conspirators largely ended, just as their bombing plot had begun.

Baghdad Steps In

Salem later complained to his FBI handlers that when he declined to carry out Nosair's plot, Nosair recruited others, like Mahmud Abu Halima. Above all, in conversations with his FBI handlers immediately after the World Trade Center bombing, Salem repeatedly reminded them of "Muhammad Salameh, with Ibrahim El Gabrowny going to visit Sayyid Nosair in jail."[11] But that did not adequately describe the full dimensions of the plot as it developed after Salem broke off contact with Nosair and El Gabrowny.

Even before the FBI dropped Salem, it seems that Baghdad had learned of the bombing conspiracy. That is what Muhammad Salameh's telephone bill suggests. Salameh was the individual who was arrested a few days after the World Trade Center bombing, as he returned to the Ryder rental agency to collect his deposit on the leased van that had been used to transport the bomb. At the same time that Nosair succeeded in convincing his cousin, Ibrahim El Gabrowny, to organize plans for a bombing, Muhammad Salameh appears to have been recruited into the plot, for he began to call El Gabrowny frequently.

Salameh comes from a long line of terrorists on his mother's side. His maternal grandfather was active in the 1936 Arab revolt against British rule in Palestine. Much later he joined the PLO and was arrested by the Israelis in the early 1980s, despite his advanced age, and died soon after his release from prison.

A maternal uncle, Kadri Abu Bakr, was number two in the PLO's "Western Sector." The Western Sector was a terrorist unit within the PLO, like Force 17 or the Hawari group. The Western Sector was established in the late 1960s, at a time when the PLO was based in Jordan, and it refers to the area west of the Jordan River: that is, Israel and the territory it captured in the 1967 war.

The Western Sector, like other PLO terrorist organizations, such as Force 17, was under Iraq's strong influence. Several members of the Western Sector are known to have carried out operations at Saddam Hussein's behest.[12]

Abu Bakr was arrested by Israeli authorities in 1968. The 1967 war had precipitated an upsurge in terrorist attacks from the newly displaced Palestinians, and Abu Bakr was among those who infiltrated from Jordan into Israel to attack targets there. He was sentenced to twenty years imprisonment, but was released in 1986, after serving eighteen years of his term, and deported from the West Bank, whence he made his way to Iraq.

In the ten days from June 7, the day after El Gabrowny met with Nosair at Attica, through June 16, Salameh called El Gabrowny seventeen times, almost twice a day. By contrast, in the previous ten days, Salameh had called El Gabrowny only four times. (See the log of Salameh's phone calls, chart 3–1.)

Salameh's frequent calls to El Gabrowny after June 6 seem to reflect Salameh's recruitment into Nosair's bombing plot. Indeed, Salem later reported that Nosair had tried to recruit Salameh at the same time that Nosair had tried to recruit Salem.

Salameh's phone bill went through the roof. It had been $128.41 in May. It was $1,401.00 in June, and $2,516.28 in July.[13] On June 16, Salameh even got the conference-calling feature for his telephone. Something had changed in Muhammad Salameh's life.

Salameh began calling Iraq. On June 10, Salameh made the first of forty-six calls to that country before his telephone service was cut off on July 9 for nonpayment. The vast majority of those calls were made to his uncle, Kadri Abu Bakr (see the page from Salameh's phone bill, illustration 3–1).

When Salameh began calling his uncle in Baghdad in June 1992, what did he tell him? Perhaps it was something like, "Oh Uncle! You are going to be so proud of me. I'm going to be like you and Grandpa. You won't believe what we're going to do. We're going to avenge martyr Nosair!" Perhaps, Salameh even asked for help in the exciting new project El Gabrowny was organizing.

CHART 3–1
LOG OF TELEPHONE CALLS FROM MUHAMMAD SALAMEH AND OF CONTACTS BETWEEN CONSPIRATORS, JUNE AND JULY 1992

June 1:	1 call to El Gabrowny
June 2:	1 call to El Gabrowny
June 5:	2 calls to El Gabrowny
June 6:	El Gabrowny visited Nosair in prison
June 7:	2 calls to El Gabrowny
June 8:	4 calls to El Gabrowny
June 9:	3 calls to El Gabrowny
June 10:	Salameh began to call his uncle, Kadri Abu Bakr, in Baghdad and spoke for 33 minutes; first of 46 calls to Iraq
June 11:	4 calls to El Gabrowny
June 14:	Salem and another co-conspirator met with Nosair in prison and discussed constructing bombs
June 16:	4 calls to El Gabrowny Salem met El Gabrowny and a co-conspirator met El Gabrowny and discussed the bombing plot
June 19:	2 calls to El Gabrowny
June 22:	1 call to El Gabrowny
June 23:	1 call to El Gabrowny
June 29:	4 calls to El Gabrowny
July 3:	1 call to El Gabrowny

SOURCE: Government Exhibit 824, *United States v. Muhammad Salameh et al.,* and Government's Memorandum of Law in Opposition to Defendants' Pretrial Motions (Phase 1), *United States v. Omar Abdel Rahman et al.*

Through Salameh's phone calls, Baghdad almost certainly learned of the New York extremists' plans. As a matter of routine, Iraqi intelligence would have had Abu Bakr's phone bugged, and Abu Bakr would have known that. The Iraqi regime would have expected Abu Bakr to tell it about something so important as a bombing conspiracy in New York. Indeed, had Abu Bakr not reported the news to Iraqi authorities, he might have raised suspicions about himself. That, in turn, would have made him liable to arrest, interrogation, and torture by Iraqi security forces. Thus, even if Abu Bakr had not been spontaneously inclined to report the activities of his nephew in New York to Iraqi intelligence, a prudent man might well have felt obliged to do so. Perhaps, even, Abu Bakr approved of his nephew's plans.

And it was not long after Salameh began calling his uncle in

ILLUSTRATION 3–1
MUHAMMAD SALAMEH'S JUNE 1992 TELEPHONE BILL

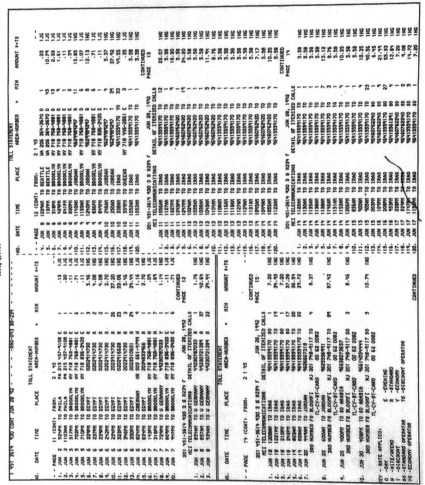

SOURCE: Government Exhibit 824, *United States v. Muhammad Salameh et al.*

Baghdad that Abdul Rahman Yasin, thirty-two years old, traveled from Iraq to Jordan. On June 21, he applied for and received an American passport from the U.S. embassy in Amman. Yasin had been living in Baghdad, but he was born in America and therefore entitled to a U.S. passport. He would later become an indicted fugitive in the Trade Center bombing.

The origins of the World Trade Center bombing would thus seem to lie with El Sayyid Nosair in Attica in May and June of 1992, and in the bombing conspiracy that arose with Emad Salem's help. Baghdad seems to have learned of the plot through Salameh's calls to his uncle. And Baghdad, apparently, decided to help out. Abdul Rahman Yasin was sent to Jordan to pick up a passport at the U.S. embassy there. From the speed of its reaction, it would seem that Iraqi intelligence was already planning such an operation. But the specific means to carry it out fell into their laps through Salameh's calls to his uncle.

It is easy to understand why the U.S. government might have trouble explaining the origins of the World Trade Center bombing in that way. But the government, for its part, offered no coherent explanation of the beginnings of the plot. Rather, it offered several contradictory explanations.

In opening arguments of the World Trade Center bombing trial, the prosecution claimed that the bombing conspiracy began in September, with the arrival in New York from Pakistan of Ramzi Yousef, the bomb's mastermind, and Ahmad Ajaj, his traveling companion. But in closing arguments, as the prosecution realized that its case against Ajaj was weak, it asserted, "The story really begins in Houston, Texas, in April 1992," when Ajaj left the United States for Pakistan, before returning in September.[14] Finally, in the second bombing conspiracy trial, the trial of Shaykh Omar et al., the government claimed that the World Trade Center bombing plot began when Yousef left Baghdad, traveling to Pakistan, before coming to America.

Why did Yousef come to New York? How did he know whom to contact? How did the foreign elements in the World Trade Center bombing interact with the Muslims resident in New York?

That is the key to understanding the bombing and who was ultimately responsible for it. To do so, we must look more closely at the accused. Who are the men who were charged with blowing up the building?

4

The Accused

Seven men were originally indicted for the World Trade Center bombing. But at the time of the first trial, which began in September 1993 and lasted until March 1994, there were only four defendants in the courtroom (see illustration 4–1). Two of the accused were fugitives, and the seventh man was tried separately on lesser charges. The four defendants in the first Trade Center bombing trial were all Muslim fundamentalists, but they were not an impressive lot. None was very clever.

But the two men who had fled successfully certainly were. They had come to the United States in September 1992 and stayed for six months. There is no particular reason to believe that either man was a Muslim extremist. In addition, a mysterious person arrived shortly before the bombing to supervise the operation in its last stages, or so authorities believed. The two original fugitives, along with the unidentified man, would seem to represent the foreign element that became involved with Shaykh Omar Abdul Rahman's followers in New York and led them in the Trade Center bombing.

The Palestinians

Three Palestinians resident in the New York area, all the same age and friends of one another, were indicted in the Trade Center

ILLUSTRATION 4–1
PHOTOGRAPHS OF THE FOUR DEFENDANTS WHO FIRST STOOD TRIAL

MUHAMMAD SALAMEH, who rented the van that carried the bomb and was arrested when he returned for his deposit on the van. Befriended by Ramzi Yousef, he lived with Yousef in New Jersey and helped him make the bomb.

MAHMUD ABU HALIMA, who helped mix chemicals for the bomb.

NIDAL AYYAD and his wife; Ayyad helped order chemicals for the bomb, and following the bombing he sent messages to news organizations claiming credit for the attack.

AHMAD AJAJ, who met Yousef in Islamabad and flew with him to JFK Airport, carrying bomb-making manuals, passports, and other material. He was arrested and detained for illegal entry. While in prison, he regularly spoke with Yousef.

SOURCES: Photographs of Salameh, Ayyad, and Abu Halima are Reuters/HO/Archive photos. Photograph of Ajaj is from the FBI.

bombing. The Palestinians were younger and far less experienced than the others involved in the bombing.

Muhammad Salameh. Salameh is the eldest son among the eleven children of a retired Jordanian warrant officer, who supplements his limited pension by working at an import clearing house. Salameh's family originally came from Bidya, a West Bank town near Nablus. They fled to Jordan shortly after the 1967 war, when Salameh was two months old. They now live in Zarqa, a dreary city 30 kilometers east of Amman, inhabited largely by Palestinian refugees.

Salameh did not grow up as an Islamic extremist. As a teenager, he played soccer and liked wrestling and Western movies. But when he tried to enter Jordan University, he could not get into the colleges of law or engineering and settled instead on studying Islamic law. Salameh dropped out of that program, however. Over his parents' objections, he traveled to New York in 1988. "He had so many dreams then," his mother explained, "of a job, of money. He told us how he was going to help us."[1]

Salameh wanted to get a degree in business administration at an American college. But his dreams bore no relation to his abilities. Salameh came on a tourist visa, and he remained in the United States illegally. Described as a loner who always needed money, he held no steady job. He had become a passionate Muslim, averting his eyes if the television was on and covering his ears when he heard the radio. He considered them to be corrupting influences. Salameh became involved in the political activities of the New York area extremists. He participated with Nosair in paramilitary activities and hung out at the local offices of the mujahedin—those fighting on behalf of Islam against the Soviets in Afghanistan—although he himself did not go there.

Salameh was also very naïve. As a neighbor commented, Salameh was laughed at by friends for being what they called "stupid." Another acquaintance remarked, "Someone could have set him up—a very strong person. . . . He's very innocent."[2] That was apparently a family trait. Salameh's family shares his politics, his hatred of Israel, and his deep sense of Palestinian grievance. But as a *Newsweek* reporter who visited Salameh's home in Jordan and met his family remarked, "There's nothing there."[3] His family was as dimwitted as he.

Nidal Ayyad. The same age as Salameh, Nidal Ayyad was Sala-meh's good friend. They had met at a Jersey City mosque, the notorious Masjid al-Salaam, a few years before. Ayyad, too, was involved in the extremists' activities, but he was far less passion-ate about them than Salameh.

Ayyad was the oldest of six children. His family moved from a West Bank village near Ramallah to Kuwait shortly after the 1967 war, and he grew up there. In 1978, his father divorced his mother and left for the United States. Ayyad was raised by an uncle, while his father settled in New Jersey, marrying an Amer-ican woman.

Ayyad was the most educated and successful of the New York extremists indicted in the Trade Center bombing. When he turned eighteen, his father brought him, his mother, and his sib-lings to New Jersey, where Ayyad began his studies at Rutgers University. His teachers described him as an average student. He graduated in January 1991 with a degree in biochemical engi-neering, and he soon got a job at Allied Signal, a New Jersey–based chemical conglomerate. Ayyad had an annual salary of $38,000 and was granted U.S. citizenship that year. He was not doing badly for a kid, a recent immigrant, just out of college.

The next year, in 1992, Ayyad decided it was time to get married. In November, his mother went to Jordan to choose a suitably religious girl for him. He was wed in Amman in Decem-ber, bringing his bride back to America and taking a new apart-ment in a quiet New Jersey suburb. By the time of the Trade Center bombing, she was pregnant.

During the sentencing of the World Trade Center defen-dants, the judge singled out Ayyad: "You are clearly the most culpable. You had the most breaks. You came to this country. You had a chance to do something with your life."[4]

Bilal al-Qaysi. The same age as Ayyad and Salameh, al-Qaysi came to the United States in 1986. He first lived in Chicago, then moved to the New York area, where he worked as a livery driver. Al-Qaysi was a friend of both Ayyad and Salameh. He shared a savings account and safe deposit box with them, related to their efforts in the Islamic cause. Not only was al-Qaysi involved in the political activities of the local fundamentalists, he had even served in Afghanistan.

Following the Trade Center bombing, al-Qaysi went to stay with friends in Carbondale, Illinois. After learning that authorities were looking for him, he surrendered himself voluntarily on March 24. Although al-Qaysi was originally indicted for the Trade Center bombing, that charge was dropped in early August, some five months after the attack. The judge was impressed by the fact that he had turned himself in. Moreover, authorities had little specific evidence against him and could not really explain his role in the bombing.

Instead, al-Qaysi pled guilty to lying to immigration officials. He was given a twenty-month prison sentence, after which he was to be deported.

The Egyptians

One Egyptian was indicted in the bombing. A second was arrested at the same time and was among the defendants who stood trial on the seditious conspiracy charges against Shaykh Omar and ten others. The Egyptians were older, tougher, and more experienced than the Palestinians.

Mahmud Abu Halima. Abu Halima was born in 1959, the first of four sons of a textile mill foreman in Kafr al-Duwwar. A town of some 250,000, twelve kilometers southeast of Alexandria, Kafr al-Duwwar is home to Egypt's second largest textile mill. Kafr al-Duwwar is remembered in Egyptian history for a labor strike in August 1952, shortly after the Egyptian army overthrew the monarchy. The workers at the mill believed that the new government heralded a new, revolutionary era, in which such things were possible. They were mistaken. The strike was crushed with a heavy hand. By the late 1970s, the town was known for its Islamic militancy.

Abu Halima became involved with outlawed Islamic extremist groups as a teen-ager. He was never arrested, but occasionally his friends were. In the spring of 1981, as tensions escalated between the extremists and the government, Abu Halima decided to leave Egypt. He had been studying for a degree in education at the University of Alexandria, and in September he quit his studies and left for West Germany.

Abu Halima sought political asylum there. But a year later,

his asylum appeal was rejected on the grounds that if, as he claimed, he had never participated in any crimes in Egypt, then he had nothing to fear from authorities. Abu Halima would have been deported, but he was saved by a sympathetic German woman, a thirty-four-year-old nurse with a history of alcoholism and emotional problems. She found him polite and friendly: "He was never violent, never aggressive," she recalls, and she was happy to marry him to block his deportation.[5]

Abu Halima assumed responsibility for providing for the both of them, working as a dishwasher, cook, kitchen helper, and so forth. He took night classes in German in order to return to school, so that he could become a teacher. He learned to speak German well, although he did not like Germans. He thought they were cold and drank too much. Instead, he found companionship within Munich's community of Muslim extremists.

When it turned out that Abu Halima's wife did not want to have children or convert to Islam, they were amicably divorced. Abu Halima married another woman, a student who dreamed of becoming a dancer, as she struggled to recover from the recent death of her seventeen-year-old brother in a motorcycle crash.

Abu Halima did not like America either. He blamed it for keeping Egypt's government in power. Still, in 1986 he and his second wife came to the United States. They remained in America illegally.

In 1990, Abu Halima received permanent resident status through the 1986 Immigration Reform Act, a fraud-plagued program intended to resolve the status of immigrant farm workers but which, in practice, was abused by many who had never been farm workers. Abu Halima, in fact, worked as a taxi driver. He garnered a string of violations for tinkering with the meter, driving with a suspended license, and so forth. A former employer described him as "very bad, a wise guy."[6]

In New York, Abu Halima was active in raising money for the Afghan cause, and he reportedly made several trips to Afghanistan. After Shaykh Omar came to the United States in July 1990, Abu Halima acted as his driver. He also helped Nosair to organize paramilitary training.

Yet whatever the extent of Abu Halima's involvement in the extremists' activities, whether in New York or in Afghanistan, he was far from being the architect of the World Trade Center bomb, as some initially thought him.[7] In fact, his first wife described him as a "technical nitwit" who did not know even how to fix a television set.[8]

Ibrahim El Gabrowny. Like his cousin El Sayyid Nosair, El Gabrowny was born in Port Said, Egypt, in 1950. Much later, in America, El Gabrowny met his wife Lisa Detweiler, an American convert to Islam, through Nosair's wife. Detweiler, a student nurse and twelve years El Gabrowny's junior, lived in Pittsburgh, while he lived in New York. Their courtship consisted of two weeks of telephone calls. They did not meet until the night before their wedding.

El Gabrowny was president of Brooklyn's Abu Bakr mosque—one of the more radical mosques in the New York area. It was dominated by Egyptian émigrés and was a center of activity against the Egyptian government. El Gabrowny was the most senior and authoritative of the New York–based extremists arrested in connection with the Trade Center bombing. Theodore Williams, who was a prisoner in the Metropolitan Correction Center (MCC) with the Trade Center defendants, came to know them and El Gabrowny well. According to Williams, "It was clear to me that El Gabrowny was [Abu Halima's] boss, but not *the* boss."[9]

El Gabrowny was very active in Nosair's defense. He quit his job as a construction contractor and went on welfare. He traveled to Saudi Arabia, where he convinced the Saudis to allow collection boxes to be placed in the mosques, raising more than $100,000 for his cousin's legal expenses. He also provided Islamic supporters, transportation, and even bodyguards for Nosair's defense team.

When Salameh rented the van that carried the Trade Center bomb, he used a New York driver's license that he had obtained in September 1992. The driver's license was in Salameh's own name, but Salameh, who lived in New Jersey and also had a New Jersey license, had used El Gabrowny's Brooklyn address to obtain the New York license.

Police went to El Gabrowny's apartment the day they arrested Salameh. They had not meant to arrest him, only to search his apartment. But they found El Gabrowny outside the building where he lived. When they attempted to search him, he resisted, assaulting two officers. He was finally subdued with the help of a third, and he was arrested because of his attack on the officers.

After his arrest, El Gabrowny was cautioned against washing his hands. He asked to use the toilet, urinated, and plunged his hands into the urine-filled bowl. That thwarted the test for explosives. However, substances on a jacket and on a piece of wood in his apartment were found to be consistent with the presence of urea and nitric acid, the two major chemicals in the Trade Center bomb.

Five Nicaraguan passports, five corresponding birth certificates, and two driver's licenses were found on El Gabrowny's person, which helps explain why he resisted the police search. The passports bore pictures of Nosair and his family, with false names. Somehow, Nosair was still hoping to escape, and El Gabrowny was helping him. But where did the Nicaraguan passports come from?

When Nicaragua's first post-Sandinista government assumed office in 1990, it discovered that 50,000 passports could not be accounted for. The left-wing Sandinistas had given nearly 1,000 passports to sympathizers from foreign countries who had taken up residence in Nicaragua. But the rest had disappeared, sometimes turning up in "the unlikeliest of places," according to a *New York Times* report. Indeed, as a Nicaraguan finance official told the *Times*, "When Iraqi troops withdrew from Kuwait in February 1991, they left behind a large quantity of Nicaraguan passports."[10]

And notably, two men came to the United States in September 1992, carrying Iraqi passports, on journeys that began in Iraq, or seemed to begin in Iraq. They were the two fugitives.

The Iraqis

Two Iraqis were instrumental in the plot, although in this chapter we shall examine only one. The second is best dealt with separately, which we will do in chapter 5.

Abdul Rahman Yasin. Yasin was born in Bloomington, Indiana, in 1960, while his father was working on his Ph.D. at the University of Indiana. Yasin grew up in Baghdad, and he came to New York in early September 1992 on a U.S. passport acquired two months earlier from the U.S. embassy in Amman. An epileptic, he ostensibly came for medical treatment, staying with his brother Musab in Jersey City. (See illustration 4–2, the wanted poster for Yasin.)

Yasin is charged with having helped mix chemicals for the bomb. That is a fairly significant role. Yasin is also a cagey fellow. Salameh was arrested on March 4. Yasin was picked up that day in a sweep of sites associated with Salameh and taken to the FBI's Newark office for questioning. But Yasin managed to pull the wool over the eyes of the FBI agents who interrogated him.

Unlike El Gabrowny, Yasin was very cooperative.[11] He provided the FBI agents with useful information. He told them he had helped Salameh learn to drive the van that had carried the bomb. And he told them about a safe house, the apartment on Pamrapo Avenue in Jersey City, which was used to mix the chemicals for the bomb.

The FBI's New Jersey office considered Yasin to be cooperative and innocent. In a sworn affidavit, Agent Eric Pilker referred to him as a confidential informant who "has no training or experience in explosives or the production of explosive devices."[12]

New Jersey FBI believed that Yasin would be helpful in the future, and they allowed him to leave their office. Rudimentary precautions were not taken. His passport was not held, although he had come from Iraq only six months before and could easily return. It also seems that airlines were not alerted that he should not be allowed to leave, as was done in the case of several others then considered suspect. The day after his questioning, Yasin flew to Jordan, and from there traveled on to Baghdad, where he was later seen by an ABC News stringer outside his father's house.

While New Jersey FBI tended to regard Yasin as a very good source of information, New York FBI took a dimmer view. Investigators subsequently tried to draw Yasin back to America, some with the aim of questioning him further, some in the belief that he should be arrested. His brother Musab was repeatedly brought to the New Jersey FBI office to telephone him in Baghdad. Musab was told to say that authorities wanted Yasin to return, just to ask him a few more questions. Yasin seemed

ILLUSTRATION 4–2
WANTED POSTER FOR ABDUL RAHMAN YASIN

$2,000,000

REWARD

Diplomatic Security Service

At approximately 12 noon on February 26, 1993, a massive explosion rocked the World Trade Center in New York City, causing millions of dollars in damage. The terrorists who bombed the World Trade Center murdered six innocent people, injured over 1,000 others, and left terrified school children trapped for hours in a smoke filled elevator.

Following the bombing, law enforcement officials obtained evidence which led to the indictments and arrests of several suspected terrorists involved in the bombing. ABDUL RAHMAN YASIN, one of those indicted, fled the United States immediately after the bombing to avoid arrest. YASIN is now a fugitive from justice. YASIN was born in the U.S., moved to Iraq during the 1960's, and returned to the U.S. in the fall of 1992. He possesses a U.S. passport. Because of the nature of the crimes for which he is charged, YASIN should be considered armed and extremely dangerous.

The United States Department of State is offering a reward of up to $2,000,000 for information leading to the apprehension and prosecution of YASIN. If you have information about YASIN or the World Trade Center bombing, contact the authorities, or the nearest U.S. embassy or consulate. In the United States, call your local office of the Federal Bureau of Investigation or 1-800-HEROES-1, or write to:

HEROES
Post Office Box 96781
Washington, D.C. 20090 – 6781
U.S.A.

ABDUL RAHMAN YASIN
DESCRIPTION

DATE OF BIRTH:	April 10, 1960
SOCIAL SECURITY NUMBER:	156-92-9858
U.S. PASSPORT NUMBER:	27082171, issued on 6/21/92 in Amman, Jordan
IRAQI PASSPORT NUMBER:	M0887925, in the name of Abdul Rahman S. Taher
PLACE OF BIRTH:	Bloomington, Indiana
HEIGHT:	5' 10"
WEIGHT:	180 pounds
BUILD:	medium
HAIR:	black
EYES:	brown
SEX:	male
RACE:	white
NATIONALITY:	Iraqi
CHARACTERISTICS:	Possible chemical burn on right thigh. Epileptic; takes medication for condition
ALIASES:	Abdul Rahman Said Yasin, Aboud Yasin, Abdul Rahman S. Taha, Abdul Rahman S. Taher

SOURCE: Department of State.

agreeable in principle; he just had a few personal matters to take care of.

Thus, even after his flight, Yasin continued to deceive the FBI, or at least some of its agents. Only after local authorities finally came to the conclusion that their efforts to lure Yasin back were fruitless did they make the decision to indict him—on August 4, 1993, some five months after the bombing.

Yasin was so smooth that it is necessary to ask whether he had been instructed in how to handle such a situation. His skillful manipulation of the FBI agents who dealt with him marks him as a cut above all the suspects discussed so far.

But there was another Iraqi involved in the plot, or at least someone who called himself an Iraqi—Ramzi Ahmed Yousef, the brains behind the bomb. The mysterious Yousef and his traveling companion, indicted and convicted for the bombing, are the focus of chapter 5.

Even today, the identity of Ramzi Yousef—who lived in America for six months; fled after the bombing, and then in February 1995 was returned to this country, where he twice stood trial; and now resides in a maximum security prison in Florence, Colorado—remains a puzzle. Ramzi Yousef is the key to unraveling the mystery of who was behind the World Trade Center bombing.

Who is Ramzi Yousef? What do we know about him? What can we learn about him?

5

Ramzi Ahmed Yousef—Mysterious Mastermind

Ramzi Ahmed Yousef was the individual most responsible for building the World Trade Center bomb. But he is a mysterious figure. Until his arrest, a point as simple as his ethnicity was unclear, and his nationality remains unknown to this day. Yousef succeeded in presenting himself as a Pakistani and obtaining a passport from the Pakistani consulate in New York, but the Arabs with whom he dealt in New York knew him as an Iraqi. "Rashid, the Iraqi," they called him. Speaking fluent English, Yousef told a human rights organization that he was an Iraqi-born victim of political persecution in postwar Kuwait. "A very nice guy, very credible—in retrospect, a very good liar," explained the paralegal who handled his case.[1]

Yousef's trip to America seems to have begun in Iraq. According to the Iraqi passport on which he entered the United States, Yousef obtained a visa for Pakistan in Baghdad in April 1992. The next month, Yousef left Iraq for Pakistan, transiting through Jordan.

But it is unclear whether the stamps on Yousef's passport are authentic. His journey may not have begun in Baghdad at all. Our first confirmed sighting of Ramzi Yousef was in the summer of 1992, when Ahmad Ajaj met him in Islamabad, Pakistan's capital city.

Yousef in Pakistan. Ahmad Ajaj, a Palestinian, was twenty-six when he first met Yousef, in 1992. Ajaj had gone to Pakistan to join the Muslim struggle there. He was largely a petty forger. In Jerusalem, he had run a currency counterfeiting operation in a cemetery, where he was arrested in 1987. Ajaj had poor machinery and was not particularly skillful, according to Israeli authorities. The next year he was arrested for document forgery in Jordan. In addition to counterfeiting money, Ajaj fabricated identity documents, mostly driver's licenses and passports, for his companions in the Islamic cause.

In early 1991, Ajaj was caught doing something more serious than counterfeiting. In May, he was deported from Israel to Jordan for a period of five years for smuggling weapons into the West Bank. In September, Ajaj left Jordan for the United States, where he sought political asylum. He moved to Houston, where an uncle lived. There, Ajaj worked delivering Domino's pizza.

But in April 1992, with his asylum application still pending, Ajaj went to Pakistan, traveling as Muhammad Gihad Abid. Abid, a twenty-six-year-old Palestinian living in Houston, had purchased a ticket to Peshawar, Pakistan, that he could neither use nor refund. He advertised at the mosque, offering to sell his ticket cheaply.

Peshawar was the base for the Muslim resistance to the Soviet occupation of Afghanistan until the Soviet withdrawal in 1989 and the fall of the indigenous Afghan Communist government three years later. Peshawar is now home to feuding Afghan factions and Muslims of various nationalities, who came for the jihad [holy war] and cannot go home because their radical politics are anathema to their own governments.

After arriving in Pakistan, Ajaj worked at the Azzam Service Center, the largest facility in Peshawar for Arab volunteers. It was founded in 1984 by a Palestinian, Abdullah Azzam. Azzam had earned a Ph.D. from Cairo's al-Azhar University, the premier theological institution in the Muslim world. He then taught Islamic law at the University of Jordan, was dismissed from that position, and moved to Saudi Arabia, before leaving for Peshawar in 1984 to establish his center. Along with two of his sons, Azzam was killed by a car bomb in 1989.

Ajaj soon discovered that he did not like the situation in Pakistan. As he would tell friends in Texas after his return to America, "I went to a country which was full of disease, of hardship, and of fatigue. By God Almighty, while you are eating, the house flies eat with you. From the excess of it, the excess of fatigue and the excess of hardship, I got sick. . . . They made for us a very hard time and drained our saliva for nothing."[2]

In July 1992, Ajaj traveled to Islamabad, 110 miles east of Peshawar, seeking a visa from the U.S. embassy that would allow him to return to America. But the embassy refused. He had failed to get permission to leave America, as he should have done with an asylum application pending. Ajaj then went to Islamabad's Arabic Center, where he discussed his problem with others. Ramzi Yousef, whom he was meeting for the first time, volunteered that he, too, was going to America, and offered to help.[3]

Indeed, Yousef was to "help" Ajaj in many ways. Part of Ajaj's problem was that he was in Pakistan illegally. On international flights the name on the ticket must match the name on the passport. Ajaj had traveled to Pakistan on Muhammad Abid's passport in order to use Abid's cheap ticket. Then, Ajaj arranged to return Abid's passport to him. But that left Ajaj in Pakistan on his own passport, without a Pakistani visa and without an entry stamp.

Ajaj's passport, as it was introduced into evidence, contains a Pakistani visa and entry stamp. But both are forgeries. Yousef probably put the false stamps into Ajaj's passport for him, making Ajaj's presence in Pakistan legal, or seemingly so.

Yousef also arranged for the purchase of two first-class tickets to New York. The tickets were paid for by voucher, obscuring the source of funds. The two went to Peshawar before flying together to New York. Thus, the tickets on which they traveled—Peshawar–Karachi–New York—made it appear that Yousef, like Ajaj, was based in Peshawar, but there has never been any independent confirmation of that.

Yousef told Ajaj and others that he had been fighting the Communists in Afghanistan. That is what Muhammad Salameh believed. But Jamal Kashoggi, who covered the war in Afghanistan for *al-Hayat*, the premier Arabic language newspaper, asked the leaders of the major mujahedin groups in Pakistan if they knew of Ramzi Yousef. None did.[4]

It is generally assumed that Ramzi Yousef was based in Peshawar and that he trained in Afghanistan and fought alongside the Muslim radicals there. But no independent evidence ever emerged to back up Yousef's claim. That he came out of the milieu of the militant fundamentalists is a critical assumption that may well be mistaken. Islamabad may have been Yousef's base.

Indeed, two years later, when Ramzi Yousef was arrested, it was in Islamabad. And the fellow who turned Yousef in—a South African Muslim named Ishtiyaak Parker—like Ajaj, first met Yousef in Islamabad.[5]

Yousef Arrives in New York. In the late summer of 1992, when Yousef and Ajaj left Pakistan for America, they left on European passports, stolen from their owners in 1991 on visits to that country. Since they were both traveling on European passports, they did not need to show visas for the United States to Pakistani Airways officials.

On September 1, Yousef and Ajaj arrived together at New York's John F. Kennedy International Airport. Yousef had only a small carry-on bag, but Ajaj had three suitcases. They were packed with a few clothes and a slew of books, including six manuals about making bombs, Molotov cocktails, and so forth, along with diagrams on how to make silencers and a videotape on suicide bombings. Ajaj's suitcases also contained two stolen passports, in addition to the one on which he traveled, as well as a variety of other identification documents belonging to young men of various Middle Eastern nationalities. These documents were precious to Ajaj, as he was a forger of identity documents, although it is unclear whether they belonged to him, Yousef, or both.

At customs, Ajaj presented the Swedish passport on which he had left Pakistan. His photo was clumsily pasted on and the deception easily detected by the scratch of a fingernail. Under questioning, Ajaj became loud and belligerent, and he was detained for illegal entry. He remained in jail until after the World Trade Center bombing.

Yousef, though, did not present the British passport, in the name of Azan Muhammad, on which he had left Pakistan. He had given that passport to Ajaj to carry, along with his ticket from Pakistan to the United States, also in the name of Azan Muhammad. Instead, Yousef showed immigration authorities his Iraqi passport, asked for asylum, and was allowed in. Authorities subsequently surmised that Yousef used Ajaj to create a diversion, facilitating his own entry into the country.

Yousef gave immigration authorities an address in Houston, but he went to stay at 34 Kensington Avenue, in Jersey City. Both Muhammad Salameh and Musab Yasin, a thirty-three-year-old Iraqi, rented apartments in that building. At roughly the same time, Musab's younger brother, Abdul Rahman, arrived from Baghdad on his American passport and came to live in Musab's apartment.

Muhammad Salameh had resided at his 34 Kensington Avenue apartment since September 1991. Many young Arab men used the two apartments, holding prayer meetings and eating large, communal meals. Relations between Musab's and Salameh's apartments were so close that an intercom was even rigged up between them. Established within this group, Yousef quickly befriended Salameh and the two soon moved out together to share an apartment elsewhere in Jersey City.

Yousef is the acknowledged mastermind of the World Trade Center bomb. Without Yousef, the New York fundamentalists could not have built the bomb. But Yousef does not fit the conventional image of a terrorist. By all accounts, he is charming, or at least he can seem so. Even though Ajaj's entry with Yousef landed him in jail for six months, Ajaj remained loyal to Yousef, as his telephone conversations, taped while Ajaj was in prison, suggest. "He's very respectable, very honest. . . . I owe him, Abu Muhammad," Ajaj explained to his friend.[6]

Ajaj, however, knew practically nothing about Ramzi Yousef. One of his concerns in prison was to help Yousef get a job. He asked a friend to help find work that would not compromise Yousef religiously, advising, "I don't expect he [Yousef] would accept work in a store that handles beer." Ajaj's friend promised to speak with someone who might have "a 'halal' [permitted] job, which does not involve forbidden things."[7]

That Yousef moved smoothly among the fundamentalists is perhaps of little surprise. But he moved with equal ease in American society. He spoke fluent Arabic, English, and Urdu, the principal language of Pakistan. He talked his way past U.S. immigration authorities, and he impressed the paralegal who was handling his supposed asylum appeal.

Yousef also pulled off some impressive telephone fraud, one more testament to his abilities. Having used a variety of long-distance carriers, he left New York owing more than $18,000 to telephone companies. A camera at an ATM machine captured him withdrawing funds and making telephone calls—on other people's cards—after his own service was cut off for nonpayment. He wore photo-gray sunglasses! As ABC's Sheila MacVicar observed, Yousef is not someone who spent the past ten years fighting for Islam in the mountains of Afghanistan.[8] The mujahedin live the life Ajaj described. Photo-gray sunglasses are not part of it.

Yousef is urbane, intelligent, and highly skilled. Indeed, that was evident when his first trial began, in June 1996, for the plot to bomb twelve U.S. airplanes. He acted as his own lawyer. To be sure, he had a court-appointed lawyer sitting at his side. And his defense was not particularly skillful. But he spoke fluent, if somewhat accented, English. And he comprehended the basic principles of U.S. law and the ebb and flow of discourse in the formal setting of a courtroom.

But who is he? The government never formally explained. As a Justice Department official once told me, "It doesn't matter what we call him. We just try a body."[9] The man was not being flippant. He was just rationalizing the position that he found himself in, because of decisions made by others.

Thus, if we want to know more about Ramzi Yousef and who he is, we must inquire for ourselves. And we can learn something about Yousef from the passport on which he left the United States. Indeed, we can learn a lot about Ramzi Yousef from that passport. It is a key piece of evidence, the significance of which U.S. authorities missed.

The Passports of Abdul Basit Mahmood Abdul Karim

Although Ramzi Yousef came to America on an Iraqi passport, in the name of Ramzi Yousef, he left on a Pakistani passport, in the name of Abdul Basit Mahmood Abdul Karim. On November 11, 1992, Yousef went to a Jersey City police station claiming to be Abdul Basit and reporting that he had lost his passport two days before. On December 31, he went to the Pakistani consulate in New York to get a new passport. Yousef presented Xerox copies of Abdul Basit's 1984 and 1988 passports. The picture on Abdul Basit's 1988 passport is similar to the picture on Yousef's 1991 Iraqi passport, the one he presented to U.S. immigration officials upon his arrival in America.

There really was an Abdul Basit Mahmood Abdul Karim. He was born in Kuwait in 1968, grew up and was educated there, graduating from high school with a concentration in science. Abdul Basit then studied in Great Britain. In the fall of 1986 he began his "A" levels, the British year of college preparation, attending a school in Oxford. The next year, he began a two-year technical course in computer-aided electronic engineering at Wales's West Glamorgan Institute of Higher Education [WGIHE], subsequently renamed the Swansea Institute. He completed what is called a Higher National Degree in June 1989 and returned to Kuwait, where he obtained a job at the National Computer Center in Kuwait's Planning Ministry.

That information, in its generalities, can be gleaned from Abdul Basit's passports. The specifics were provided by veteran British journalist Tony Geraghty, a former defense correspondent for the *Sunday Times of London*. At the time, late 1994 and early 1995, I was working with ABC's Nightline. Nightline hired Geraghty to do a little sniffing around. Geraghty did an outstanding job, and his work became especially important after Ramzi Yousef's arrest, as we shall see in chapter 16. Also, Tim Phillips, a reporter for the Swansea Institute's student newspaper, subsequently provided some information.

Yet, is the person who masterminded the World Trade Center bombing the same person as Abdul Basit Mahmood Abdul Karim, the engineering student? That is: Is the picture on the 1988 passport that of the same person whose picture appears on the 1984 passport? (See appendixes A and B, Abdul Basit's 1984 and 1988 passports.)

Part of the problem is adolescence. The 1984 passport shows a smooth-faced boy of sixteen, the 1988 passport a bearded young man of twenty. The 1988 passport gives Abdul Basit's height as thirty centimeters taller than the 1984 passport, more than a foot. Two different people, or an adolescent growth spurt?

The 1984 passport gives a permanent address in Karachi, the 1988 passport an address in Turbat, in Pakistani Baluchistan, near the Iranian border. Throughout, Abdul Basit's family was resident in Kuwait. Why should they change their address? In fact, the address on a Pakistani passport should not change. It refers to a family's place of origin, which need not necessarily be their place of residence.

And if Yousef really were Abdul Basit, Yousef would have been twenty-four years old at the time of the bombing, the youngest member of the conspiracy. Was the youngest member of the plot likely to have been the mastermind?

U.S. immigration took Yousef's picture upon his arrival in September 1992. That picture is shown in illustration 5–1. Does it look like the face of a twenty-four-year-old? Or does it look like someone older? Ajaj, Yousef's traveling companion, thought that Yousef was in his thirties.

Additionally, the signatures of Abdul Basit beneath his pictures in the 1984 and 1988 passports are quite different. Even someone unfamiliar with Arabic script, in which Urdu is written, should be able to see that. The signature on the 1984 passport is a proper signature in Arabic script. The signature on the 1988 passport is not. It is a scrawl in what appears to be undecipherable Latin letters. (See appendixes A and B.)

Ramzi Yousef did not like to sign his name in America, particularly in Arabic script. U.S. authorities have no document with his signature in Arabic. Notice that the Iraqi passport on which he entered the United States is unsigned (see illustration 5–2). So, too, is his application, as Abdul Basit, for a new Pakistani passport (see illustration 5–3). Why? Did Abdul Basit have one signature and Ramzi Yousef another?

The Pakistani consulate was not fully satisfied with the documentation presented by Ramzi Yousef. They gave him a six-month passport, on the understanding that he would get a regular passport once he returned "home" to Pakistan.

A skillful forger and a good Xerox machine, or computer-cum-scanner, can radically change a document. It seems that both the 1984 and 1988 Abdul Basit passports are genuine, but

ILLUSTRATION 5–1
WANTED POSTER FOR RAMZI YOUSEF

$2,000,000

REWARD

At approximately 12 noon on February 26, 1993, a massive explosion rocked the World Trade Center in New York City, causing millions of dollars in damage. The terrorists who bombed the World Trade Center murdered six innocent people, injured over 1,000 others, and left terrified school children trapped for hours in smoke filled elevators.

Following the bombing, law enforcement officials obtained evidence which led to the indictments and arrests of several suspected terrorists involved in the bombing. RAMZI AHMED YOUSEF, one of those indicted, fled the United States immediately after the bombing to avoid arrest. YOUSEF is now a fugitive from justice. YOUSEF was born in Iraq or Kuwait, possesses Iraqi and Pakistani passports, and also claims to be a citizen of the United Arab Emirates. Because of the nature of the crimes for which he is charged, YOUSEF should be considered armed and extremely dangerous.

The United States Department of State is offering a reward of up to $2,000,000 for information leading to the apprehension and prosecution of YOUSEF. If you have information about YOUSEF or the World Trade Center bombing, contact the authorities, or the nearest U.S. embassy or consulate. In the United States, call your local office of the Federal Bureau of Investigation or 1-800-HEROES1, or write to:

HEROES
Post Office Box 96781
Washington, D.C. 20090 – 6781
U.S.A.

RAMZI AHMED YOUSEF

DESCRIPTION

DATE OF BIRTH:	May 20, 1967 and/or April 27, 1968
PLACE OF BIRTH:	Iraq, Kuwait, or United Arab Emirates
HEIGHT:	6'
WEIGHT:	180 pounds
BUILD:	medium
HAIR:	brown
EYES:	brown
COMPLEXION:	olive
SEX:	male
RACE:	white
CHARACTERISTICS:	sometimes is clean shaven
ALIASES:	Ramzi A. Yousef, Ramzi Ahmad Yousef, Ramzi Yousef, Ramzi Yousef Ahmad, Ramzi Yousef Ahmed, Rasheed Yousef, Rashid Rashid, Rashed, Kamal Ibraham, Kamal Abraham, Abraham Kamal, Muhammad Azan, Khurram Khan, Abdul Basit.

SOURCE: Department of State.

ILLUSTRATION 5–2
SIGNATURE PAGE FROM RAMZI YOUSEF'S IRAQI PASSPORT

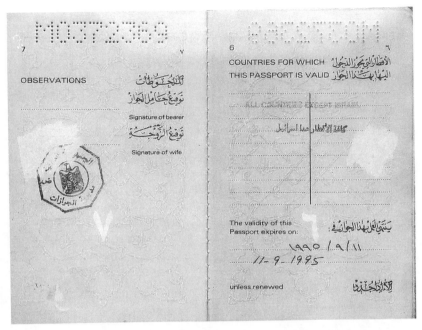

SOURCE: Government Exhibit 614, *United States v. Muhammad Salameh et al.*

items in the Xerox copy of the 1988 passport—picture, signature, and permanent residence—were changed to accommodate Ramzi Yousef, his real identity and his physical features.

In addition to the signature, Abdul Basit's height is a problem. Let us consider it carefully, as recorded in various years, in a variety of documents:

- In October 1984, Abdul Basit was sixteen years old. His height was 1 meter, 40 centimeters, or 4′7″, according to his Pakistani passport.
- In December 1988, Abdul Basit was twenty years old. His height was 1 meter, 70 centimeters, or 5′8″, according to his Pakistani passport. Thus, he grew 13 inches between the ages of sixteen and twenty.

ILLUSTRATION 5–3
RAMZI YOUSEF'S APPLICATION FOR A PAKISTANI PASSPORT

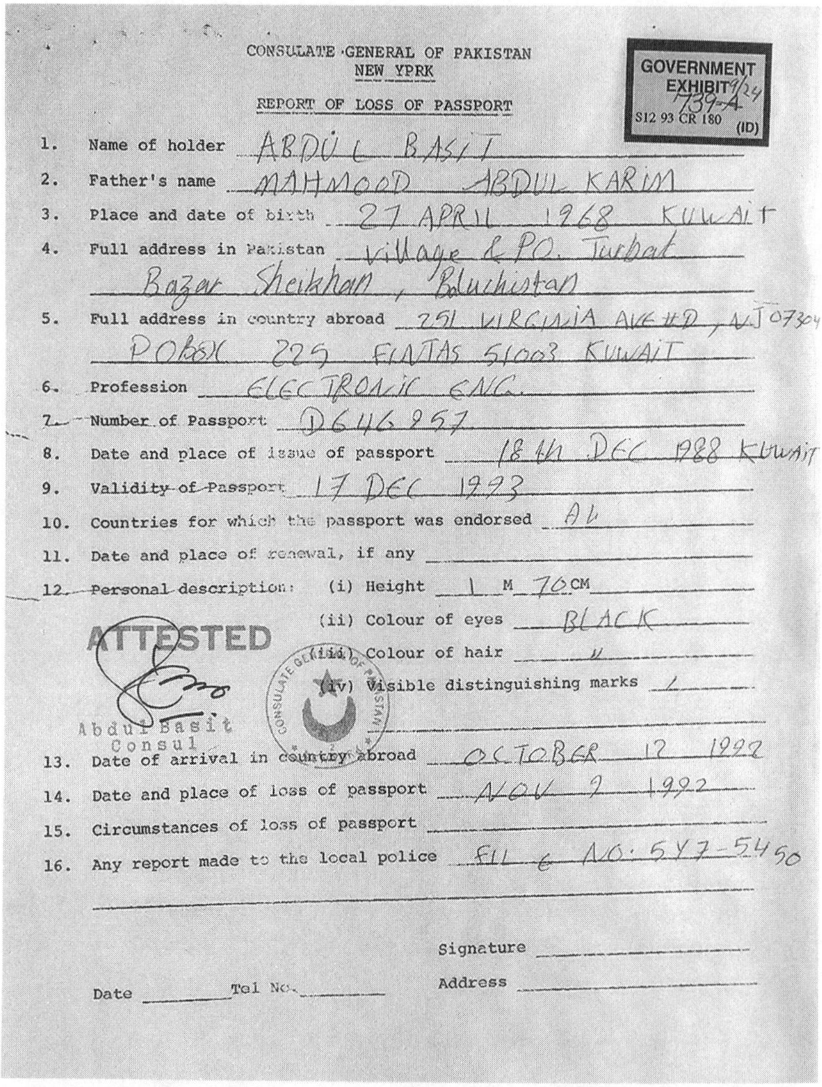

CONSULATE GENERAL OF PAKISTAN
NEW YORK

REPORT OF LOSS OF PASSPORT

GOVERNMENT
EXHIBIT
739-A
S12 93 CR 180 (ID)

1. Name of holder   ABDUL BASIT

2. Father's name   MAHMOOD ABDUL KARIM

3. Place and date of birth   27 APRIL 1968 KUWAIT

4. Full address in Pakistan   Village & P.O. Turbat Bazar Sheikhan, Bluchistan

5. Full address in country abroad   251 VIRGINIA AVE #D, NJ 07304 P OBOX 225 FINTAS 51003 KUWAIT

6. Profession   ELECTRONIC ENG.

7. Number of Passport   D646857

8. Date and place of issue of passport   18th DEC 1988 KUWAIT

9. Validity of Passport   17 DEC 1993

10. Countries for which the passport was endorsed   AL

11. Date and place of renewal, if any

12. Personal description: (i) Height   1 M 70 CM

                           (ii) Colour of eyes   BLACK

                           (iii) Colour of hair   "

                           (iv) Visible distinguishing marks   /

ATTESTED

Abdul Basit
Consul

13. Date of arrival in country abroad   OCTOBER 12 1992

14. Date and place of loss of passport   NOV 9 1992

15. Circumstances of loss of passport

16. Any report made to the local police   FIL & NO. 547-5450

Signature

Date   Tel No.   Address

SOURCE: Government Exhibit 739A, *United States v. Muhammad Salameh et al.*

- In September 1991, Abdul Basit was twenty-three years old. His height was 1 meter, 75 centimeters, or 5′9″, according to Ramzi Yousef's Iraqi passport. Thus, Abdul Basit grew 1 inch between the ages of twenty and twenty-three.
- In September 1992, Abdul Basit was twenty-four years old. Ramzi Yousef was 6′ tall, according to U.S. immigration. Thus, Abdul Basit grew 3 inches in one year, between the ages of twenty-three and twenty-four.

Could anyone have grown at that pace? Even if we set aside the Iraqi passport as unreliable and concentrate only on the Pakistani passports, how likely is it that Abdul Basit, who was 4′7″ when he was sixteen and 5′8″ when he was twenty, could have been 6 feet tall when he was twenty-four?

The information regarding Abdul Basit's height on the 1988 passport is correct. Tony Geraghty learned from a source who knew Abdul Basit at WGIHE that Abdul Basit was 5′8″.[10] Boys generally stop growing around the age of eighteen, adding, on average, 1 inch to their height after that.[11] That is because sex hormones stop growth, causing the bones to fuse. According to Dr. Robert Ratner, an endocrinologist at George Washington University Medical School, a fully bearded twenty-year-old young man would not grow significantly, unless there were some abnormality in pubertal development.[12]

Abdul Basit was 5′8′. Ramzi Yousef was 6′, according to U.S. immigration. It looks like Abdul Basit is one person and Ramzi Yousef is another.

Notably, the last official mark on Abdul Basit's 1988 passport is a notation made on May 22, 1990, by the Pakistani embassy in Kuwait, changing the designation of his profession from "student" to "electronic engineering."

Iraq invaded Kuwait two months later. About 95,000 Pakistanis lived in Kuwait then. Some two dozen Pakistani nationals were seized and taken hostage by Iraqi troops.[13]

What happened to Abdul Basit and his family? Abdul Basit's father worked for Kuwait Airways. After Kuwait's liberation, he should have returned to Kuwait to ask for his old job back. But he did not. Nor can Abdul Basit or his parents be found in any other country.

Abdul Basit and his parents had no security record in Kuwait, nor did Abdul Basit have any criminal record in Great Britain. They seem to have been a typical middle-class family. If they cannot be found anywhere, there is a good chance that they are dead, killed during Iraq's invasion and occupation of Kuwait.

Abdul Basit and his parents could even have been among the two dozen Pakistanis taken hostage by Iraq. They could have died by accident, and they could have died by design. In any event, Abdul Basit's passports could easily have fallen into the hands of Iraqi intelligence. Certainly, Iraqi intelligence viewed the occupation of Kuwait as a unique opportunity to acquire identity documents. Even when they did not arrest or kill individuals, they confiscated passports and other documents from those trapped in Kuwait.

And it would seem from Abdul Basit's passports, particularly the 1988 passport, that the Iraqis did take them from him. Abdul Basit seems to have been in Kuwait in August 1990, when Iraq took over the country. Above all, the 1988 passport is incomplete. To understand why the passport is incomplete and why that is significant, we must examine the passports carefully. (See chart 5–1, "Reading Abdul Basit's Passports.")

The Missing Pages from Abdul Basit's Passports—
What They Reveal

A Pakistani passport contains thirty-six pages. Ramzi Yousef gave the Pakistani consulate twenty-one pages from the 1984 passport, but he presented only nine pages from the 1988 passport. The information contained in the 1988 passport cannot be explained on the basis of the material he gave the consulate. Pages are missing.

The last exit or entry stamp in the 1988 passport is a June 22, 1989, departure stamp from Great Britain. But the last stamp in a passport cannot long be an exit stamp. When Abdul Basit arrived at his destination, an immigration officer would have stamped an entry stamp into his passport. According to an official at Swansea Institute, Abdul Basit returned to Kuwait in June 1989, after completing his studies.[14] But no entry stamp to Kuwait, or anywhere else, follows the British exit stamp in Abdul Basit's passport.

The only item on the 1988 passport subsequent to the June 22, 1989, exit stamp is the May 22, 1990, notation, made by the Pakistani embassy in Kuwait, changing Abdul Basit's profession. Where is the entry stamp into Kuwait that would account for Abdul Basit's presence in Kuwait then? It is missing.

Indeed, more than an entry stamp into Kuwait is missing. According to the stamps in Abdul Basit's passports, as presented to the Pakistani consulate by Ramzi Yousef, Abdul Basit returned to Kuwait for Christmas vacation on December 14, 1988. He renewed his passport at the Pakistani embassy in Kuwait, as it was to expire in February 1989, when he would be in Great Britain. He obtained a new passport, while retaining possession of his old one.

Abdul Basit returned to Great Britain on December 31. He traveled in the spring, leaving England on March 16 and returning on April 7, probably on Easter vacation. But where did he go? Immigration stamps from the country to which he traveled are missing. (See chart 5–1: specifically, "The 1988 Passport: What Ramzi Yousef Withheld.")

Indeed, more than that is missing. Once we realize which stamps are missing from Abdul Basit's passport, we can understand why they were withheld.

The June 22, 1989, U.K. departure stamp on the 1988 passport marked Abdul Basit's return to Kuwait after finishing his studies. A Kuwaiti residency visa should follow, as the visa on Abdul Basit's 1984 passport expired in October 1989. Abdul Basit would have had to renew that visa on his 1988 passport.

In sum, the missing pages in Abdul Basit's passport should contain a Kuwaiti residency visa, a Kuwaiti entry stamp, and more immigration stamps from Kuwait or another country.

The Pakistani consul general in New York stated that the consulate passed every page of the passports that Ramzi Yousef gave them on to U.S. officials.[15] A spokesman for the U.S. attorney's office confirmed that the government did not withhold any pages from the passports.[16] Only Ramzi Yousef, or his sponsor, could have withheld those pages. But why?

Dangerous and Incriminating Evidence

It seems that the last stamp on Abdul Basit's passport is an entry stamp into Kuwait. He was in Kuwait at the time of the Iraqi invasion.

CHART 5–1
READING ABDUL BASIT'S PASSPORTS

THE 1984 PASSPORT

Issued: October 17, 1984, Embassy of Pakistan, Kuwait
Born: Kuwait, April 27, 1968
Profession: Student
Height: 1 meter, 40 centimeters
Permanent Address: K. 12–45 Near Jama Masjid Usmamia, Jehangir Quart., Karachi
Observation: The holder of this passport previously traveled on passport number A 51363, dated March 1, 1984, issued at Kuwait [that is, his mother's passport]

10/21/84:	Kuwaiti residency permit, valid until 10/20/89
5/11/85:	Saudi visa issued by the Saudi Embassy in Kuwait
5/16/85:	Departure from Kuwait
26 Shaban 1495:	Arrival in Saudi Arabia, according to the Muslim calendar
5/23/85:	Entry into Kuwait
11/16/86:	Visa issued by the British Embassy in Kuwait
11/17/86:	Departure from Kuwait
11/17/86:	Entry into England
8/6/87:	U.K. stamp, "The holder is exempt from requiring a visa if returning to the United Kingdom to resume earlier leave before 11/17/87."
8/10/87:	Departure from England
8/10/87:	Entry into Kuwait
9/9/87:	Departure from Kuwait
9/9/87:	Entry into England
9/22/87:	Stamp from South Wales Constabulary, Aliens Department, Cockett Police Station, Swansea West
U.K. stamp:	"Holder is exempt from requiring a visa if returning to the United Kingdom to resume earlier leave before 10/31/89."
6/20/88:	Departure from England
6/20/88:	Transit visa through Yugoslavia, and Belgrade immigration stamp.
6/21/88:	Entry into Kuwait

9/6/88:	Departure from Kuwait
9/6/88:	Entry into England
9/19/88:	Stamp from Cockett Police Station, Swansea West, South Wales Constabulary, Aliens Department
12/13/88:	Departure from England
12/14/88:	Entry into Kuwait
12/30/88:	Departure from Kuwait

THE 1988 PASSPORT

Issued:	December 18, 1988, Embassy of Pakistan, Kuwait
Height:	1 meter, 70 centimeters
Address:	Village and P.O., Turbat, Bazar Sheikhan, Baluchistan
Observations:	The holder of the Passport previously traveled on Passport A 582993, dated 10/17/84, issued in Kuwait, which has been canceled / reported lost / is attached bearing valid visa and returned to the holder.
5/22/90:	Profession has been changed at page 1 to read "Electronic Engineering" [by] Mohammed Akhtar Khan, Second Secretary, Embassy of Pakistan in Kuwait.
12/31/88:	Entry into England
U.K. stamp:	"Holder is exempt from requiring a visa if returning to the United Kingdom to resume earlier leave before 10/31/89, as endorsed in previous passport."
3/16/89:	Departure from England
4/7/89:	Entry into England
6/22/89:	Departure from England

THE 1988 PASSPORT, PAGES 10–36:
WHAT RAMZI YOUSEF WITHHELD

3/16/89:	Entry stamp into?
4/7/89:	Departure stamp from?
6/22/89:	Entry stamp into Kuwait
Fall 1989:	Kuwaiti residence visa

SOURCE: Author.

If Yousef had presented the Pakistani consulate with the full 1988 passport, an alert consular officer might have recognized that the passport seemed to have been taken from its owner during Iraq's occupation of Kuwait. Only by adding a false exit stamp from Kuwait and a forged entry stamp into another country could that danger have been avoided. Given that such forgeries ran the risk of detection, it may have seemed easier, and no more risky, just to withhold the back pages of the passport.

That would explain how Abdul Basit's passports, or at least copies of them, came into Ramzi Yousef's possession, and why he withheld the back pages of the 1988 passport. Iraqi occupation forces took the passports from Abdul Basit while he was in Kuwait. If Ramzi Yousef had presented the copies of the passports in their entirety, that could have created major problems.

Abdul Basit's complete 1988 passport would have suggested Iraqi sponsorship of the World Trade Center bombing after the event, and could have been dangerous for Yousef himself before the event. If someone had recognized that the copies of Abdul Basit's passports were copies of passports seized in Kuwait during the Iraqi occupation, that would have cast suspicion on Ramzi Yousef while he was still in America. What was he doing with such documents? Was he, in fact, working for the Iraqis?

Imagine if some Pakistani consular official had reported to the FBI the following: "We have someone who is asking for a new passport, but he has strange documentation. He is giving us Xerox copies of passports that seem to have been seized from a Pakistani citizen during the Iraqi occupation of Kuwait."

Imagine that Pakistani consular official adding, as he spoke to the FBI, "The documentation is odd, because people who have lost passports don't usually keep Xerox copies of their current passport, along with a Xerox copy of their old passport. Usually, they present some other form of identification, like a driver's license or a birth certificate. We think this is something you should look into." Indeed, a U.S. counterterrorism official readily acknowledged that such documentation as Yousef presented would immediately be regarded by U.S. authorities as a "fraud indicator."[17]

As it turned out, the danger of Yousef's being discovered through Abdul Basit's passports was remote. The Pakistani consulate was sloppy. Even the temporary passport issued to Yousef should not have been given him. Still, Yousef was very careful. He did not want to take the risk, however improbable, that the Pakistani consulate in New York would discover how he obtained copies of Abdul Basit's passports. Hence, he withheld the back pages of the passport.

There is another mystery about the Abdul Basit passports that further suggests that Yousef came into possession of them as part of a larger intelligence operation. Yousef did not have the Abdul Basit passports in his possession when he entered the United States. Someone gave or sent them to him. Who? Authorities do not know.[18]

Ramzi Yousef and Abdul Basit are almost certainly two different people. Their signatures are different, and it is very difficult to reconcile the difference in Abdul Basit's height, as given in his 1984 and even his 1988 passports, with Ramzi Yousef's height, as recorded by U.S. immigration in September 1992. Furthermore, it is a strain to believe that the picture taken by U.S. immigration of Ramzi Yousef is that of a twenty-four-year-old. Finally, it is improbable that anyone who expected to topple New York's tallest tower onto its twin would leave the United States in his own name, particularly if he had to go to some effort, as Ramzi Yousef did, to obtain the passport.

Yet, in the period before Yousef's arrest, there was a division of opinion. Some U.S. authorities thought that Ramzi Yousef and Abdul Basit were the same person, that the real identity of the World Trade Center bomber was Abdul Basit. They should have asked the question posed by Israel's savvy ambassador to the United States, Itamar Rabinovich, who began his career in military intelligence studying Syria: "Why did Yousef take the risk of going to the Pakistani consulate with such flimsy documents?"[19]

It was intended that U.S. authorities would ultimately conclude that Yousef was in fact Abdul Basit. Then, among other things, authorities would stop trying to determine his real identity. In fact, it is standard procedure for Soviet-style intelligence agencies to develop a false identity for agents involved in such missions.

Thus, it would seem that Iraq tampered with Abdul Basit's file in Kuwait while it occupied that country to generate a false identity for Ramzi Yousef. Key documents are missing from

Abdul Basit's file. Above all, copies of the front pages of his passport should be there. They are gone. Kuwait has no photograph
of Abdul Basit.

Also, there is information in Abdul Basit's file that should
not be there. Most significantly, there is a notation explaining
that Abdul Basit and his family left Kuwait on August 26, 1990,
traveling to Iraq. They crossed from Iraq into Iran at Salamchah
(near Basra) on their way to Baluchistan, where they live now.[20]

Who put that information into Abdul Basit's file? And why?
The Kuwaiti government had ceased to exist. Iraq was an occupation authority, bent on establishing control over a hostile population, amid near universal condemnation, as an American-led
coalition threatened war. The situation was chaotic. Hundreds of
thousands of people in Kuwait and Iraq were fleeing for their
very lives. Iraq took some nationals hostage, restricted the movements of others, and daily changed the rules about who could
leave from where.

Third-world nationals got the rawest deal. Westerners were
valued pawns in a high-stakes international drama. But no one
cared much about those from the third world. Tens of thousands
remained trapped for months in a desert wasteland, just inside
Jordan, because the Jordanian government would not let them
into the country, even for transit.

Was an Iraqi bureaucrat really sitting in Kuwait's Interior
Ministry recording the flight plans, including the itinerary and
final destination, of all those individuals fleeing Kuwait? That is
doubtful. Rather, it looks as if Iraqi intelligence put that information into Abdul Basit's file to make it appear that Abdul Basit
had left Kuwait, not died there, and that he had the same ethnicity as Yousef, namely Baluch, as we shall see in the next chapter.

The reason why some U.S. authorities believed that Ramzi
Yousef's real identity was Abdul Basit is that Yousef's fingerprints are in Abdul Basit's file in Kuwait. U.S. authorities had
Yousef's fingerprints, which were taken in September 1992,
when he was briefly detained for illegal immigration. (See the
INS card, illustration 5–4.) After U.S. authorities determined
that Yousef had left America as Abdul Basit, they sent the fingerprints to Kuwait, asking if they matched. When Kuwaiti

ILLUSTRATION 5–4
RAMZI YOUSEF'S FINGERPRINTS

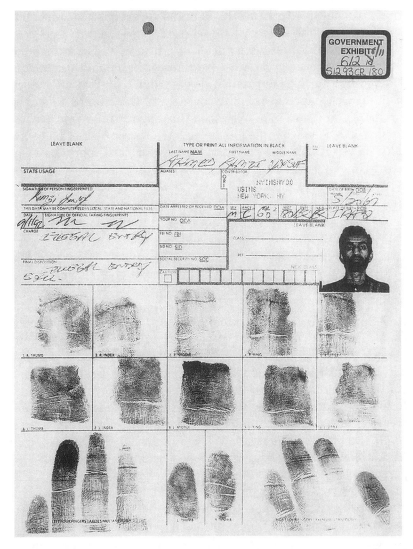

SOURCE: Government Exhibit 612, *United States v. Muhammad Salameh et al.*

authorities replied "Yes," it seemed that Yousef was Abdul Basit Karim.

Fingerprints are critical for investigators. Clerks err in recording height. Eyewitnesses blunder over photographs, but "fingerprints don't lie," or so investigators are taught.

But Kuwait's files were not secure. Iraq was in occupation of the country for seven months. Iraqi forces hauled off practically everything that was not nailed down. They took Kuwait's stock of passports, paper for making passports, the stamps to stamp passports, and so forth. Why not the fingerprint cards too? It would have been a simple matter to have Ramzi Yousef place his inky paws on a blank Kuwaiti fingerprint card, perhaps even using Kuwait's own ink, and then switch the cards.

That does not mean that Iraq was planning the World Trade Center bombing that early. It only suggests that Iraqi intelligence saw the occupation of Kuwait as a unique opportunity to develop alternate identities for some key agents.

That would also explain why Ramzi Yousef came into the United States as he did and left as he did. He came in such a way as to give U.S. authorities his fingerprints, and he left in such a way that his prints would eventually be found in a file in Kuwait. And that, it was intended, would cause U.S. authorities to conclude that Yousef's real identity was Abdul Basit, so they would cease their efforts to learn who, in fact, he really is.

U.S. authorities learned a lot about Abdul Basit from Kuwaiti and British authorities. To the little they knew about Ramzi Yousef—his height, weight, photograph, and fingerprints—they added an immense amount of information about Abdul Basit; or, at least, those who believed that Yousef was Abdul Basit did so. And they began looking for a composite person.

Basic steps to determine the real identity of the man who blew up the World Trade Center were not and have not been taken, even now, after Yousef has been twice tried and twice sentenced to spend the rest of his life in prison. Yet the question of Yousef's real identity is critical.

The Decisive Evidence against Iraq?

That Ramzi Yousef is one person and Abdul Basit another should constitute the key evidence against Iraq. Because fingerprints are so decisive—no two individuals' match—it is necessary to ask, "How did Ramzi Yousef's fingerprints get into Abdul Basit's file in Kuwait, and how was that file otherwise tampered with?"

The very fact that fingerprints are so decisive can be flipped around and turned against Iraq. By the legal standard of reasonable doubt, only Iraq could reasonably have known of or caused the death of Abdul Basit and his family and tampered with Kuwait's Interior Ministry files, above all switching the fingerprint cards. Of course, Kuwaiti authorities could have tampered with their files. But no reasonable person would suggest that Kuwait was behind Yousef's terror. Thus, only Iraq fits the bill.

And the situation is fraught with irony, unintended consequence. In the attempt to conceal themselves, the Iraqis reveal themselves. They were too clever by half.

If Ramzi Yousef is not Abdul Basit Mahmood Abdul Karim, who is he? Clues can be found in his apparent escape route.

6

Baluchistan—The Escape Route

When Ramzi Yousef went to get a new passport from the Pakistani consulate on December 31, 1992, he seems to have already worked out his escape route, which, in turn, was linked to the 1988 Abdul Basit passport, which identified him as a Baluch. That is what Yousef's telephone calls suggest.

Starting on December 3, Yousef began to make conference calls to Iranshahr, Iran, and other cities along what looks to be his escape route—by air from New York to Quetta, Pakistan, provincial capital of Pakistani Baluchistan, via Karachi; then across the border into Iranian Baluchistan; south to the coast of the Gulf of Oman; and then on to Oman. (See the map of Baluchistan, illustration 6–1.)

Yousef made many calls to Iran. A cursory glance at his telephone bill suggests an Iranian link to the bombing. But every single call Yousef made to Iran was to Iranian Baluchistan, an area Tehran does not control. In fact, the area is notorious for smuggling, particularly drug smuggling. In late 1992, a clash occurred between smugglers and Iranian Revolutionary Guards. The smugglers prevailed, taking a number of Revolutionary Guards prisoner.

ILLUSTRATION 6–1

MAP OF BALUCHISTAN—RAMZI YOUSEF'S SUSPECTED ESCAPE ROUTE

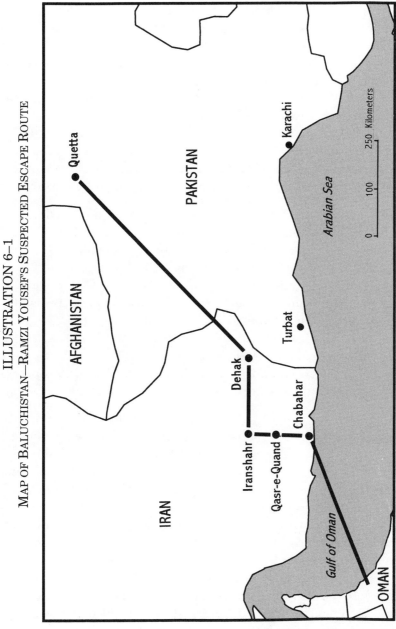

SOURCE: Author.

The telephone number in Iran that Yousef called most frequently belongs to a house in Iranshahr, in Iranian Baluchistan. The house has a frequent turnover of tenants. The people who lived there in the fall of 1992 had already left by the spring of 1994. Indeed, they had probably left much earlier, as it seems they helped Yousef arrange his escape.

As the *Washington Post* explained in the context of another dramatic terrorist case, "Pakistan's porous western border with Iran allows almost unhindered travel between Quetta and several Iranian border towns about 400 miles away across desert."[1] Yousef made the following calls in December 1992, before applying for a Pakistani passport—calls that suggest his escape route:

- 12/3/92   Conference call to Iranshahr and Oman. The Omani number belongs to a Baluch.
- 12/4 and 12/6/92   Conference calls to Iranshahr and Karachi. The Karachi number belongs to a Baluch.
- 12/15/92   Conference call to Iranshahr and Qasr-e-Qand, a town in Iranian Baluchistan, midway between Iranshahr and the Gulf of Oman.
- 12/20/92   Conference call to Iranshahr and Konarak, a town in Iranian Baluchistan, on the Gulf of Oman.
- 12/27/92   Conference call to Quetta and Dehak, a town in Iranian Baluchistan, close to the Pakistani border.

When Yousef left the United States, he flew to Karachi, with an onward flight to Quetta, Pakistan. The phone calls suggest that from Quetta he went west to Dehak, crossing the Pakistani-Iranian border, then continued further west to Iranshahr, and from there traveled south, along a major highway through Qasr-e-Qand, to Konarak, on the coast, near Chah Bahr. Chah Bahr is a large port, which has considerable trade with Oman. In Oman, the telephone trail ends. From Oman, Yousef could have gone anywhere, whether by land, sea, or air.

Baluchistan is remote and desolate. "Most observers compare it to the moon, but a team of U.S. Geological Survey experts on assignment in Iranian Baluchistan insisted that it was the closest thing on earth to Mars."[2] Before the Muslim conquest of the area, an Arab officer sent to obtain intelligence on the region reported,

> Commander of the Faithful! It is a country of which the
> mountains are mountains indeed; . . . [I]t is a country
> with so little water that its dates are the worst of dates,
> and the inhabitants are the most warlike of men.[3]

Baluchistan is so desolate that strangers who are not themselves
Baluch would be noticed traveling alone—not something anyone
who had just blown up the World Trade Center would want.

It seems that Ramzi Yousef is either Baluch or is familiar
with the Baluch, and that he had access to a number of people
resident in Baluchistan who were ready to help him with his
journey, although they were unlikely to have known the reason
for his trip.

Yousef had one persona in America, that of an Iraqi Arab,
and another for his escape, that of a Pakistani Baluch. His ability
to radically shift his identity contributed to his successful flight
and subsequently threw American investigators off his trail.
After all, who would look for an Arab in Baluchistan? Yet it would
seem that Yousef is either one or the other, an Arab familiar with
the Baluch or a Baluch familiar with Arabs.

Who are the Baluch, and what is their connection to the
Arabs?

The Baluch, the Arabs, and Iraq

There are some 3 to 5 million Baluch. They speak a language
related to Persian. Their territory is slightly larger than France,
but they are a people without a country. Their land is divided
among Pakistan, Iran, and Afghanistan.

Historically, there has been considerable exchange between
Baluchistan and the Arab countries of the Persian Gulf littoral.
In Oman, the Baluch long served as the sultan's guards, and the
oil boom of the early 1970s precipitated significant Baluch migration to other Gulf Arab countries. Conversely, there has long
been Arab migration to the coastal regions of Baluchistan. Most
notably, Oman controlled the port of Gwadar, in Pakistani Baluchistan near the Iranian border, until 1958.

Ethnic tensions surround the Baluch in Iran, Pakistan, and
Afghanistan, where they are everywhere a minority. The Baluch
are Sunni, and in Shi'i Iran religious tensions exist as well. In
fact, the Baluch have long been at odds with Tehran. In the twen-

tieth century, they were subject to an aggressive "Persianizing" policy by Reza Shah and then by his son, Muhammad Reza Shah, America's strategic partner in protecting the oil of the Persian Gulf until his overthrow in 1979.

Tehran kept Baluchistan backward and underdeveloped. Literacy rates in Iranian Baluchistan are extremely low. Twenty years ago, there were estimated to be only some 350 Iranian Baluch college graduates and another 5,000 high school graduates. Traditionally, the Iranian Baluch looked to the Pakistani Baluch for leadership, as the latter enjoyed more political and cultural freedom, making for a higher degree of political awareness and activity among them. The advent of the Islamic Republic only sharpened sectarian tensions between Tehran and the Baluch.

According to one Baluch legend, the Baluch originally came from Aleppo, in present-day Syria. That idea facilitated Arab dealings with the Baluch, on the grounds that the Baluch were really Arab.[4] The Iranian Baluch were supported in the 1960s and 1970s by left-wing, nationalist Arab governments, including Egypt's Gamal Abdel Nasser and the Ba'thist regimes in Iraq and Syria. The Baluch also received support from the Palestine Liberation Organization (PLO). A Baluch leader, Jumma Khan, was a member of the PLO's central advisory committee in the 1960s, and in 1973 George Habash's Popular Front for the Liberation of Palestine (PFLP) trained some forty Baluch fighters in Beirut.

Jumma Khan was born in Turbat, the same city listed as Ramzi Yousef's permanent address on the passport with which he fled. Turbat is in southwestern Pakistan, 125 kilometers from the Iranian border. Turbat is the major political and economic center in Pakistan's Makran region.

The Makran, the first set of Baluchistan's hills and valleys inland from the Arabian Sea, extends all the way from central Pakistan in the east to the Iranian port of Bandar Abbas in the west. The Makran is an important area of Baluch settlement, because it is one of the few Baluch areas that can support agriculture. That activity, along with remittances from workers in the Arab states of the Persian Gulf, has created an economic base that allows for a more politically active and intellectually sophisticated existence than the relatively primitive, subsistence life of the nomads of the desolate desert regions. In the Makran there exist wealthy, powerful, and politicized Baluch families.

Turbat was a major center of Baluch revolt, against both the Iranian and the Pakistani governments. In 1964, Jumma Khan established the Baluchistan Liberation Front. The front depended on support from radical Arabs and, in turn, adopted a position in solidarity with "our Arab brethren" in their struggle against "imperialism, colonialism, and Zionism."[5]

Jumma Khan took up residence in Damascus in the 1960s, but because of intense pressure for his extradition, he fled to Baghdad in 1968, where he soon became a significant figure in an Iranian Baluch revolt. Baghdad supported a Baluch rebellion against Iran from 1969 to 1973, when the shah succeeded in enticing a major tribal leader to take up a very comfortable residence in Tehran, dealing a severe blow to the Baluch opposition. Iraqi support for the Baluch continued at a reduced level until the 1975 Algiers agreement between Saddam and the shah, in which they agreed to settle their disputed boundary and to end the support each had been giving to the other's ethnic opponents. The shah sold out Iraq's Kurds, while Saddam sold out Iran's Baluch.

Still, Iraqi ties with the Baluch did not end completely. Iraq kept up contacts with Baluch Front leaders, many of whom fled to Persian Gulf shaykhdoms.

In a development related to the Baluch revolt in Iran, there was also a Baluch revolt in Pakistan, from 1973 to 1977. Turbat played a prominent role in that rebellion also. So, too, did the Marri, a powerful tribe in northeastern Pakistani Baluchistan. For at least two centuries, since the time of the British presence in the region, the Marri, the largest Baluch tribe, were fierce opponents of those foreigners who sought to control their territory.

The immediate precipitant to the 1973 Pakistani Baluch revolt was the dismissal of Baluchistan's provincial government early in the year by Zulfiqar Ali Bhutto, then prime minister and father of the subsequent prime minister, Benazir. In February 1973, Pakistani authorities discovered some 300 Soviet submachine guns and 48,000 rounds of ammunition, meant for the Baluch. They claimed to have found the weapons in the residence of the Iraqi military attache in Islamabad.[6] Three days later the attache disappeared, to be executed by the Iraqi regime several months later.

Sher Muhammad Marri, a militant Baluch nationalist who

favored a strategy of armed struggle against Islamabad, had visited Baghdad the year before. There he had struck a deal with the Iraqi government for the shipment of arms, to be shared between his group and the Iranian Baluch. He was seemingly betrayed by a rival Baluch leader, who tipped off the Pakistani government.

Bhutto dismissed the recently established provincial government of Baluchistan at the same time that Pakistani authorities disclosed their discovery of the illicit arms cache. Bhutto also asserted that the Baluch were in league with Iraq and the Soviet Union in a plot to dismember Pakistan and Iran.

Bhutto's move was, in the first place, an assault on the Baluch and their honor. A traditional people, they responded in a traditional manner. The Baluch tribes began ambushing Pakistani army convoys. Bhutto immediately went to Tehran, still fighting its own Baluch rebellion. The shah told Bhutto that he would provide Pakistan some $200 million in emergency aid to fight the Baluch.

Returning to Islamabad, Bhutto sent considerably more Pakistani troops to Baluchistan. He also arrested and imprisoned three of the most prominent Baluch political leaders: Gaus Bux Bizenjo, governor of Baluchistan; Ataullah Mengal, chief minister in the Baluchistan Government and head of the Mengal tribe, the largest Baluch tribe after the Marri; and Khair Bux Marri, chief of the Marri tribe and chairman of the National Awami [People's] Party, which governed Baluchistan province.

When the Indian subcontinent was divided between Hindu India and Muslim Pakistan in 1947, the Baluch had opposed incorporation into Pakistan. They had wanted their own state. Occasional, intermittent conflict ensued. But the fighting that began in 1973 was of an entirely different order. It approached civil war. At its height, more than 80,000 Pakistani troops were involved in suppressing the Baluch.

Helicopters proved key. The Pakistani army first used the "relatively clumsy" Chinook helicopters it had received from the United States.[7] But in mid-1974, Iran sent thirty of its U.S.-supplied Huey Cobras, replete with Iranian pilots. The Cobra had been developed for counterinsurgency use in Vietnam. Its auxiliary equipment included a six-barrel, 20 millimeter automatic cannon, which fired 750 rounds per minute.

Like other nomadic peoples, the Baluch move with their livestock to grazing areas in the summer. In the summer of 1974, the Marri men were in the hills, fighting the Pakistanis. The women, children, older men, and animals were camped in the green and fertile Chamalang Valley, in northeastern Baluchistan. On September 3, Pakistani troops attacked them in a calculated bid to bring the tribesmen out of the hills.

The ploy worked. The Marri men came to the defense of their families. For three days, Pakistani forces rained artillery fire on the encampment, occasionally strafing it with F-86 and Mirage fighter planes, along with the Huey Cobras.

The battle at Chamalang marked the turning point in the Pakistani-Baluch war. It was the beginning of the end of the Baluch revolt. In late 1975, as they continued to lose ground, some Baluch went to Afghanistan, where the royalist government of Daud Khan allowed them to take refuge. The Baluch revolt simmered on until 1977, when Zia ul-Haq ousted Bhutto in a military coup and then hanged him. Zia released some 6,000 Baluch from prison, along with the three Baluch leaders who had been under arrest since 1973.

In effect, the Baluch revolt was over. Yet some hard-line nationalists, particularly the Marri, remained in Afghanistan until 1992. Shortly before his assassination in 1978, in a Soviet-backed coup that first brought the Afghan Communists to power, Daud Khan had promised Islamabad that he would return the Baluch guerrillas. However, the new Communist government immediately granted them political asylum and formally recognized their Baluch Liberation Front.

Najibullah, chief of Afghanistan's internal security apparatus and later the country's president, even gave land to some Marri that had been abandoned by Pashtun, the dominant Afghan ethnic group, who had joined the mujahedin resistance against Kabul. The Marri remained in Afghanistan until the winter of 1992–1993, when they returned to the Quetta area, following the overthrow of Najibullah in April and the end of communism in Afghanistan. The Islamic ouster of Najibullah put an end, at least temporarily, to the ties between Kabul and the Pakistani Baluch.

Although the Baluch issue in Pakistan remained quiescent after 1977, that was not true of the Baluch question in Iran. With the 1978–1979 Iranian revolution, followed by the Soviet invasion of Afghanistan in December 1979, Arab governments thought to use the Iranian Baluch against both the Ayatollah Khomeyni and the Soviets. Arab intellectuals began to support Baluch independence. In 1979 an Iraqi author, Ma'n al-Hakkami, published a 310-page book that argued that the Baluch were Arab and made a very strong case for Arab sponsorship of an independent Baluchistan. Reportedly, Jumma Khan also appeared in Baghdad soon after the shah's overthrow.[8]

But significant Arab support for the Iranian Baluch did not materialize then, because of Islamabad's sensitivity to its own Baluch revolt. Indeed, immediately after the Soviet invasion of Afghanistan, the nightmare scenario was that Moscow would succeed in detaching Pakistani Baluchistan, thereby gaining access through Afghanistan and Baluchistan to the Gulf of Oman. Islamabad was still quite prickly about its own Baluch problem, and Arab governments deferred to Pakistani sensibilities then. Yet, over time, as Pakistan's Baluch question remained quiet, Islamabad became less anxious about them, and Arab governments, particularly Iraq, began to support the Baluch.

During the Iran-Iraq war, which lasted from 1980 to 1988, and especially in its latter stages, when Iraq was on the defensive and it even looked like Iraq might lose, Baghdad tried to create major problems for the Iranian government in eastern Iran, in Baluchistan. Iraq worked with the Baluch in an effort to divert Tehran's resources away from the war front in the west to the other side of the country.

General Wafiq Samarrai headed Iraqi military intelligence during the Iran-Iraq war. He was removed from that position in June 1991, as Saddam began to suspect his loyalties, shifting him to the palace, where he could be better watched. In late 1994, Samarrai defected to the Iraqi opposition.

Samarrai explained that Iraqi intelligence indeed has deep and well-established contacts with the Baluch on both sides of the Pakistan-Iran border. During the war with Iran, Iraqi military intelligence even maintained an office in the Arab emirate of Dubai, whence it ran Baluch as spies into Iran.

Presently, Baluchistan is a major route for transporting Iraqi products, such as oil and pesticides, into Pakistan, in violation of the UN economic embargo on Iraq. In December 1993,

Pakistani authorities announced that they had confiscated a tanker in Quetta containing 10,000 liters of Iraqi oil. The owner of the tanker was Iranian, and the driver was a Pakistani who smuggled goods to and from Iraq. A spokesman for the Pakistani federal police explained the difficulty in controlling such trade—the population on both sides of the border had a strong sense of solidarity and, as the region was so poor, the Baluch relied on smuggling for their livelihood.[9]

Given Ramzi Yousef's ties to the Baluch—as evidenced by his apparent escape route and his flight from the United States as a Pakistani Baluch—the Baluch aspect of the World Trade Center bombing constitutes another significant point suggesting an Iraqi role, as Baghdad maintained such extensive contacts with the Baluch.

From a careful analysis of the evidence presented in the first Trade Center bombing trial, it was possible to conclude that Yousef was either a Baluch with ties to Arabs or an Arab with ties to the Baluch. Something in his background allowed him readily to switch ethnic identities.

With Ramzi Yousef's arrest in February 1995, it became clear that he was indeed Baluch. When he was seized, he was carrying a Pakistani identity card in the name of Ali Muhammad Baluch, identifying him as a resident of Pasni, on the Arabian Sea, some 100 kilometers southeast of Turbat. The Pakistani press immediately reported the close ties between Saddam Hussein and the Baluch and suggested that Ramzi Yousef was working for Iraqi intelligence.[10] And when Yousef appeared in the courtroom, he indeed looked Baluch. He was dark, tall, and wiry, distinctive features of the Baluch.

And Pakistan's Baluch are, for the most part, not religious extremists. They are "notably casual about religious observances."[11] Their political history is bound up with secular nationalism and left-wing causes, including communism. Describing a meeting with Pakistani Baluch student leaders in 1978, after the first Communist coup in Kabul, Selig Harrison explained, they "spoke enthusiastically to me about the 'genuinely progressive' and 'national' character of the Kabul government and its leaders." After the Soviet invasion of Afghanistan, the students' greatest regret seemed to be that Islamabad would be able to exploit the Soviet action to its advantage.[12] That sentiment suggests how remote the perspective of the young Baluch intellectuals was from that of the Islamic radicals. For the latter, the invasion of a Muslim land by a non-Muslim power was cause for war. For the left-wing Baluch, it was secondary.

The source of the information that Ramzi Yousef is a Muslim extremist who long fought in Afghanistan is Ramzi Yousef himself. That claim, however, remains unconfirmed by any independent source. Given Yousef's ethnic background, a Baluch, it is far from evident that he is a religious extremist. In fact, with Yousef's arrest and then his trial for the airplane bombing conspiracy, it became clear that Yousef is definitely not a radical Muslim, as we will discuss in chapter 15. If Yousef ever spent any time in Afghanistan, he would have been in Najibullah's camp rather than with the Islamic mujahedin.

Did U.S. authorities recognize or understand the significance of the Baluch aspect of the World Trade Center bombing? In a major court document, the government asserted that Yousef fled to Peshawar.[13] But Yousef's ticket, presented as evidence in the first Trade Center trial, gave Quetta as his destination, "UET" in airline code. Yousef fled to Quetta, not Peshawar. (See Yousef's ticket, illustration 6–2.)

ILLUSTRATION 6–2
AIRLINE TICKET WITH WHICH RAMZI YOUSEF FLED

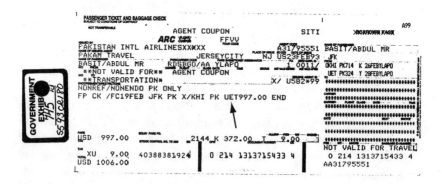

SOURCE: Government Exhibit 745, *United States v. Muhammad Salameh et al.*

Yousef's destination in Pakistan is extremely significant, because from Pakistan he went on to a third country. Peshawar is fast on the Afghan border, and a thousand miles from Iran. If Yousef went to Peshawar, that implies an escape into Afghanistan, not Iran. But from Quetta, Yousef could easily have crossed into Iran. And Yousef's escape route has implications for the questions, Who was behind him? and What did he represent?

If Yousef had fled to Afghanistan, that would be consistent with the government's claim that he arose out of the milieu of the mujahedin there. If he fled through Baluchistan, however, it would suggest that he had little or nothing to do with the Muslim extremists in Afghanistan. Rather, at a moment of extreme personal danger, Yousef chose not to be with them, but in Baluchistan. And, of course, Baghdad has well-established ties with the Baluch.

7

The World Trade Center Bombing Operation—The Bomb Itself

The World Trade Center bomb was no ordinary bomb. It was the most ambitious terrorist attack ever, anywhere, meant to cause hundreds of thousands to die. Following his arrest in Pakistan, the mastermind, Ramzi Yousef, told authorities escorting him back to the United States that he had hoped to kill as many as 250,000 people.[1] Thus, New York was extremely fortunate that the fatalities were few. Still, the bomb sent more people to hospitals than any other single incident on a single day in America's peacetime history.

The Force and Composition of the Bomb

The conspirators intended to topple the North Tower onto the South Tower, or so FBI engineers concluded after studying the placement of the bomb-laden van, which had been left parked against the support columns of the North Tower. And that is what Yousef himself told authorities after his arrest.

The World Trade Center bomb might have brought down another skyscraper, like Chicago's Sears Tower, but it could not have caused the collapse of the Trade Center. Building codes in the Northeast are stricter than they are in the rest of the country, and it is much more difficult to down the Trade Center. The towers were constructed to extremely rigorous standards, with "enormously redundant" structural support, according to their designer, Leslie Robertson, who assisted in the World Trade Center reconstruction.[2] Once the world's tallest buildings, now the third tallest, the twin towers were meant to withstand the impact of hurricane force winds, a terrorist attack, or a 707 jet plane crashing into them. But the towers sit on a large landfill, essentially a mud basin, although the bombers could not have known that. The bomb came close to cracking the wall that contains the mud fill.[3]

The immense destructiveness of the World Trade Center bomb is generally not recognized, because the overwhelming preponderance of the damage was below ground, in the basement floors. (See illustration 7-1.) The bomb, placed on the B-2 level of the North Tower's parking garage, created a crater six stories high. The upward force of the blast tore through ceilings of steel and reinforced concrete 11 inches thick on each of two levels above the bomb, leaving an 18 x 22–foot hole in the lobby of the Vista Hotel, which lies between the two towers. The downward force of the blast turned the floor of the B-2 level in the area beneath the bomb into powder, obliterating 15,000 square feet of steel and concrete. The blast, along with 2,500 tons of falling debris, punched a hole through the floor of the B-3 level, as well as the floor of the B-4 level, below.

At the base of the crater, level B-5, burning debris piled up three stories high. The force of the explosion cracked a 5-foot-diameter cast-iron pipe that brought water from the Hudson River to the building's air conditioning chillers. Some 1.8 million gallons of water flowed into the basement, shorting out emergency back-up generators on B-6, the only basement level otherwise untouched by the blast.

The explosion scattered material at a rate close to 20,000 feet per second. A 7-ton steel diagonal support brace from the North Tower was ripped away and thrown 40 feet. It slammed into a brick wall, causing the wall to collapse onto five maintenance workers who were eating lunch.

ILLUSTRATION 7–1
DIAGRAM OF WORLD TRADE CENTER TOWERS

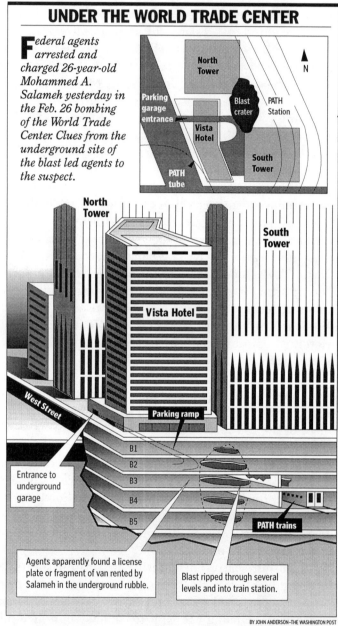

UNDER THE WORLD TRADE CENTER

Federal agents arrested and charged 26-year-old Mohammed A. Salameh yesterday in the Feb. 26 bombing of the World Trade Center. Clues from the underground site of the blast led agents to the suspect.

North Tower

Parking garage entrance

Vista Hotel

PATH tube

Blast crater

PATH Station

South Tower

N

North Tower

South Tower

Vista Hotel

West Street

Parking ramp

B1
B2
B3
B4
B5

Entrance to underground garage

PATH trains

Agents apparently found a license plate or fragment of van rented by Salameh in the underground rubble.

Blast ripped through several levels and into train station.

BY JOHN ANDERSON–THE WASHINGTON POST

SOURCE: © 1993 *The Washington Post*. Reprinted with permission.

In the immediate wake of the blast, authorities worried about "progressive collapse," the possibility that one load-bearing column might fall, shifting its weight onto other columns, collapsing them in turn. Two-foot-thick concrete casements surrounding three vertical support columns had been blown away, leaving the steel girders bare, unsupported for 50 feet. The basement floors of the building served as horizontal supports, but five of them had huge, gaping holes.

Material for repairing the building was fabricated overnight. Emergency workers drilled holes 4 feet in diameter into the floor of the Vista Hotel, through which steel girders were inserted below. Welders suspended in cages were lowered down to secure the beams and prevent the building's collapse. That "only" six people died from such a blast is amazing. It was not just luck, but also testimony to the skill of those who designed and built the towers.

Indeed, Jim Fox, who as director of the New York office of the FBI supervised the bombing investigation, described that crater to me. He explained what it was like to put in a hard, long, dusty day of work, searching for evidence amid the debris. At the end of the day he would sometimes sit, relaxing with a beer, at the edge of the crater. And he would ask himself, "Who the hell did this?"[4]

Those trying to carry out terrorism in the United States face a difficulty they do not encounter in other places, like the Middle East. There, explosives are readily available on the black market. In the United States, explosives are tightly regulated. The World Trade Center bombers overcame that problem by concocting a home brew in an apartment that Yousef and Salameh rented as a safe house. The apartment, at 40 Pamrapo Avenue, Jersey City, was set well back from the street, affording the seclusion from prying eyes that they needed to create the bomb, undetected.

Making the World Trade Center bomb was a difficult and dangerous task. The plotters needed great skill to avoid blowing themselves up. They used nitroglycerin as a trigger, making their own. Nitroglycerin is highly unstable and explodes easily. Changes in temperature or a slight shock, like the scuff of a shoe on a small amount of material spilled on the floor, will set it off.

The bomb's main explosive component was urea nitrate—1,200 pounds of it, more than half a ton. Urea nitrate is a crystalline substance, made by mixing urea and nitric acid. But urea nitrate is fairly stable. It does not explode easily. Hence, the need for the nitroglycerin trigger.

The bomb consisted of the mass of urea nitrate, divided up into manageable weights, wrapped in plastic garbage bags, and packed into cardboard boxes. Nitroglycerine, suspended in cool water, was fixed into the urea nitrate. The water served to insulate the nitroglycerine from any changes in temperature or shocks on the drive to Manhattan. The explosive compound was given an extra punch by the addition of several "boosters": lead azide, magnesium, and ferric oxide. Then, three canisters of hydrogen gas were placed across the boxes of the urea nitrate-based explosive to give the bomb an extra punch.

If all that were not enough, the bomb had yet one more alarming feature: those trapped in the North Tower were to be asphyxiated in a cloud of cyanide gas. That was first explained publicly by Judge Duffy during the May 1994 sentencing of the first set of Trade Center bombing defendants. Addressing the defendants, Duffy said, "Death is what you really sought to cause. You had sodium cyanide around and I'm sure it was in the bomb. Thank God the sodium cyanide burned instead of vaporizing. If the sodium cyanide had vaporized it is clear that what would have happened is that cyanide gas would have been sucked into the North Tower and everyone in the North Tower would have been killed."[5]

Judge Duffy's account of the conspirators' intent to produce cyanide gas is supported by official reports of the materials found in the locker in which the conspirators stored the chemicals for the bomb. Among them were sulfuric acid and sodium cyanide. An FBI forensic chemist, Steven Burmeister, testified during the trial that sodium cyanide is not used as an explosive but, when mixed with sulfuric acid, it produces hydrogen cyanide. "You breathe it, you're dead."[6]

Is it reasonable to suppose that this most ambitious terrorist bombing ever was solely the work of émigré Muslim extremists in New York? Could they have conceived of such a plot—toppling one tower onto another, amid a cloud of cyanide gas—let alone executed it? Israeli intelligence believed that the bomb had a foreign sponsor, and pointed to Iran—the foremost opponent of the peace process, and the country with which Israel was also engaged in regular conflict, in Lebanon.[7]

Iraq, however, is most known for the use of poison gas. Before the Gulf War, Iraq had the largest, most diversified chemical weapons program in the third world. Iraq used chemical weapons against Iran during the Iran-Iraq war, and against its own Kurdish population in the same period. During its war with Iran, Iraq became the first country in history to use highly toxic nerve agents on the battlefield, as opposed to the relatively primitive, less effectual poison gases used in World War I.

Historically, the most notorious use of hydrogen cyanide was, of course, Adolf Hitler's effort to exterminate European Jewry. Early Nazi attempts to kill significant numbers of people by simply shooting them caused too many psychological problems for those charged with the task. So following Hitler's decision in 1941 to wipe out the Jewish population in Nazi-controlled territory, German authorities experimented with different methods. They first tried carbon monoxide, which they had used earlier in eugenics projects. But they found carbon monoxide ineffectual for killing on the scale they envisaged. Sometimes the victims only fell unconscious and did not die. The Nazis resolved their problem when they hit upon Zyklon-B, a gas used for fumigating against rodents and insects. Zyklon-B was the German trade name for hydrogen cyanide. The Nazis produced hydrogen cyanide in the same way that the World Trade Center bombers were said to have produced it—by mixing acid with a cyanide compound such as sodium cyanide.

There is also a link between Saddam's chemical weapons program and sodium cyanide. After the autumn of 1983, when Iraq began to use chemical weapons in its war with Iran, Western industrialized countries imposed export controls on the chemical precursors of such weapons. By 1987, at Australia's initiative, twenty states and the European Commission had committed themselves to enforcing such restrictions. Nevertheless, in 1991, the British government reported that a British company had during the preceding year exported more than one million dollars' worth of sodium cyanide to Iraq, in contravention of the Australia Group's guidelines.

At the same time, Iraq was also trying to acquire the technology to make sodium cyanide out of locally available material. An American entrepreneur, Louis Champon, had developed a unique process for producing bitter almond oil from the pits of apricots and peaches. The oil gives foodstuffs such as cherry soda and cherry vanilla ice cream their cherry flavor. Champon touted his bitter almond oil as the only "natural" cherry flavoring; he sought investors to help him set up a processing plant for his new product. A London-based, Iraqi-born businessman, Ihsan Barbouti, was ready to put $5.3 million into Champon's business, and a plant was soon established in the Florida resort town of Boca Raton, not far from Miami. It began operating in September 1989.[8]

Small amounts of cyanide are found naturally in apricot and peach pits. Champon's process produced, as a by-product, ferric ferrocyanide, a chemical used as a metal cleaner and found in such substances as Drano. It is basically harmless and can be disposed of by flushing down the drain. Barbouti proposed shipping the chemical by-product to Germany, but Champon rejected the notion.

Barbouti appointed as plant manager a chemist who Champon later learned was an expert in the production of hydrogen cyanide, and who had supervised the construction of a hydrogen cyanide manufacturing plant in Louisiana for the Swiss chemical conglomerate Ciba Geigy. The plant manager proposed a slight change in Champon's plans for processing the pits, claiming that it would make production cheaper. Unknown to Champon the result was to produce sodium cyanide as a by-product of the fruit pit processing, instead of the harmless chemical envisaged in Champon's original plan.

U.S. authorities later discovered that sometime soon after Champon's plant started operations, seven barrels of sodium cyanide, some 1,500 pounds' worth, were shipped abroad. The cyanide was trucked from Florida to Houston—where Barbouti had a corporate headquarters—and then on to Baltimore, from where it was shipped to Europe, labeled as the "personal effects" of a diplomat at the Iraqi Embassy in Washington, and therefore exempt from U.S. customs inspection.[9] From Europe, the cyanide was shipped to the Jordanian port of Aqaba, and on by road to Iraq. Authorities surmised that the limited quantity of sodium cyanide was shipped to Iraq principally to demonstrate that Champon's process actually worked.

No one really knows, however, what happened to the rest of the sodium cyanide produced at Champon's plant. It was supposed to be flushed away, as if it were the original by-product, ferric ferrocyanide. But Champon later wondered if Barbouti had not installed a storage tank to catch the sodium cyanide, as it was seemingly disposed of, and thus if much more sodium cyanide did not find its way to Iraq.[10]

There is no doubt that Barbouti had close links with the Iraqi regime. He had been a consultant to the Iraqi Defense Ministry, and his acquisition efforts closely matched Baghdad's needs. Shortly before he invested in Champon's enterprise, Barbouti sought to obtain another sensitive technology in America.

Moshe Tal, an Israeli who emigrated to the United States in the early 1970s, was something of a chemistry genius. His company, TK-7, produced chemical mixtures that, when added to ordinary fuels, gave them an extra kick—in the same way that chemical compounds were used to give an extra punch to the basic explosive material in the World Trade Center bomb. Tal's hottest product, developed with an eye toward military markets, boosted the power of several fuels, including liquid rocket fuel. In March 1987, Barbouti approached Tal about buying the company.

In 1987, Iraq seemed closer than ever to losing its war with Iran. In January, Baghdad faced Tehran's most threatening assault of the conflict. The Iranians made critical gains around the southern Iraqi port city of Basra, which remained under siege for three months. Notably, the offensive came on the heels of the November 1986 revelation of the covert U.S.-Israeli arms sales to Tehran. The Iranian leadership sought to recover from its embarrassment; to exploit Iraq's confusion about American intentions; and to utilize the weapons it had received in the arms-for-hostages deals. In the spring of 1987, Iraq dramatically accelerated its military procurement and development programs. Saddam's cousin and son-in-law, Hussein Kamil—who defected in

1995 and then was killed when he returned to Baghdad six months later—was made supervisor of military industries.

Known as "the Whip," Kamil brought with him to military industries men from Saddam's personal bodyguard: uneducated, coarse, and ignorant youths from the Sunni Arab towns around Tikrit, Saddam's home base and the bastion of his support. Kamil imposed a six-day workweek and twelve-hour workday on the military factories. To maintain an accelerated pace, the guards beat employees for offenses like sleeping on the job or leaving their machines. They also turned the guards' room into a brothel, forcing themselves on female workers.[11]

Among Kamil's priorities was to increase the range of Iraq's Soviet-made, liquid-fueled Scud missiles. Baghdad wanted to strike Tehran, some 450 miles from the Iran-Iraq frontier, but no missiles existed in the Iraqi arsenal with sufficient range. Moshe Tal's rocket-fuel booster would be very useful. Saddam's half-brother, Sabawi Ibrahim al-Tikriti, was among those present at a London meeting in the fall of 1987 during which Barbouti's takeover of TK-7 was discussed.[12] Sabawi at that time played a key role in Saddam's overseas procurement program, before becoming head in 1989 of Iraqi General Security, responsible for maintaining internal control.

In September 1987, Moshe Tal had a meeting scheduled with Barbouti in Zurich, from where the two were to fly together to London. Barbouti was late. Tal called Barbouti's London office and was told that Barbouti's flight had been delayed and he should check his time of arrival—from Tripoli—with Libyan Airways. To Tal, an Israeli, this was a red flag.

On their flight to London, Tal plied Barbouti with liquor and learned Barbouti's real business. Among other things, Barbouti told Tal that he was involved in the construction of a chemical weapons plant in Rabta, Libya. He also explained that he was involved in an effort to build a heavy water plant in Libya, from which nuclear material could be extracted. Colonel Qaddafi wanted to smuggle an atomic device into U.S. waters on a commercial ship and then blow up a port city, in retaliation for America's 1986 raid on Tripoli.[13] Of course, Qaddafi never came remotely close; he settled for bombing a U.S. airliner, Pan Am 103.

Tal went to both Israeli and U.S. authorities with his information. It was soon confirmed that Barbouti was the principal procurer of equipment for Libya's chemical weapons factory. Many of the supplies for the plant were coming from Germany, and the Reagan administration pressured Bonn to cut off the flow. The Germans resisted, and in frustration Washington, in January 1989, publicly identified Barbouti as the "linchpin" in the construction of the Rabta plant. At the same time, Belgian authorities arrested one of Barbouti's associates. The next year, another Barbouti partner, from the German chemical company Ihmausen Chemie, was arrested, tried, and sentenced to five years in prison by a German court for his role in building Libya's chemical weapons factory. In June 1990, German authorities issued a warrant for Barbouti's arrest. But two weeks later, on July 1, 1990, he died suddenly of a heart attack, at the age of sixty-three. Twenty-one years before, Barbouti had faked his own death to cover his tracks. Had he done so again? The inscription on Barbouti's "tombstone" in the English countryside, in Arabic and English, reads, "His life a beautiful memory, His absence a silent grief. Forever in our thoughts."[14]

Washington was slow to understand and accept the conclusion reached by New York law enforcement officials that a cyanide gas attack was intended to accompany the bombing of the World Trade Center. Indeed, in December 1994 an official from the CIA's Counterterrorism Center scoffed at the notion, dismissing Judge Duffy's charge as so much "grandstanding."[15] But five years on, the claim is generally accepted. Indeed, Secretary of Defense William Cohen, in a *Washington Post* article discussed in the introduction to this book, stated just that.[16] But Cohen used the poison gas dimension of the Trade Center bombing to argue that the United States faced a serious danger from a new terrorist source—"individuals and independent groups." Yet the World Trade Center bombing does not demonstrate that at all. Rather, it looks very much to be terrorism of the old variety—state-sponsored.

8

The Structure of the Conspiracy

Rather remarkably, the World Trade Center bomb was built while the FBI itself was investigating the Muslim extremists in New York, including one of the World Trade Center conspirators. But authorities missed the bombing plot, because (1) they were focused on the wrong people; (2) supervisors in Washington did not take the matter seriously; and (3) the World Trade Center bombing was a compartmentalized operation.

The Compartmentalization of the Operation—Avoiding the FBI

After the FBI dropped its informant, Emad Salem, in July 1992, it began an investigation of the Egyptian extremists in New York, prompted, at least in part, by information that Salem had provided. Two different federal groups were involved in the investigations—one an FBI terrorism task force, the other a group seeking to build a civil rights case upon which to try Nosair for Kahane's murder. Shaykh Omar Abdul Rahman, El Sayyid Nosair, and Mahmud Abu Halima were among those targeted for investigation.

The Egyptian government was consulted over the course of the investigation. "There were many, many contacts between

Cairo and Washington," explained a high-ranking Egyptian official.[1] But American and Egyptian authorities held two different opinions of Shaykh Omar and his followers. The Egyptians maintained that the shaykh's mosques were dangerous hotbeds of terrorist activity, while the Americans took a less alarmist view.

There was, in fact, some tension in American-Egyptian relations then, over this issue. In addition to Egyptian suspicions that the United States had allowed Shaykh Omar into the country as a hedge against a radical Islamic takeover in Cairo, Egypt also believed that Shaykh Omar's haven in America was payback for his participation in the CIA-backed Muslim campaign against the Soviets in Afghanistan.

Indeed, the government has never provided a fully satisfactory explanation of how Shaykh Omar was repeatedly allowed to travel to and from America, and to take up residence here. A report by the State Department's inspector general, Sherman Funk, concluded that the seemingly suspect approvals of Shaykh Omar's travels to and residence in America were merely repeated and multiple bureaucratic errors.[2] Funk's conclusion may well be warranted, but those errors were considerable, and it is understandable why that explanation might have been met with some skepticism.[3]

Shaykh Omar gave religious sanction for the 1981 assassination of Anwar Sadat, telling the Muslim extremists who sought his approval that the act was indeed lawful.[4] Nonetheless, an Egyptian court acquitted him of that charge in 1984. Subsequently, he was arrested repeatedly and imprisoned for activities such as inciting violence and attacking police officers. In August 1987, Shaykh Omar was placed on a State Department terrorist watch list.

Still, in May 1990 Shaykh Omar received a multiple entry visa, valid for one year, from the U.S. embassy in Khartoum, and he came to America in July. The consular officer in charge was a CIA agent, working undercover, and not particularly interested in who received visas to the United States. A local Sudanese employee actually approved Shaykh Omar's visa application, and the employee may well have worked for Sudanese intelligence.

But the mistakes that allowed Shaykh Omar repeated access to America did not end there. On November 5, the day Kahane was assassinated, the shaykh was in London. He again entered the United States in mid-November, without any problem.

Kahane's assassination brought Shaykh Omar's presence in the United States to public attention, and the State Department revoked his visa at the end of November. On December 16, the *New York Times* reported: "Islamic Leader on U.S. Terrorist List Is in Brooklyn." But that very day, Shaykh Omar reentered the United States, this time after a trip to Denmark. On January 31, 1991, Shaykh Omar applied to the Immigration and Naturalization Services for permanent residency. Although obtaining permanent residency is usually a time-consuming process, INS granted it to Shaykh Omar after only two months. It was little wonder that Cairo was edgy.

As it turned out, U.S. authorities were essentially correct about the capabilities of the émigré Islamic extremists in New York; they were limited. Egyptian authorities were correct about their proclivities; they were more than ready to act upon the inflammatory rhetoric of Shaykh Omar and, indeed, others like him. What was missing in the American-Egyptian exchanges was the understanding that a third party might enter the scene to bridge the gap between the extremists' abilities and their aspirations, to turn wild talk into lethal action.

Both Egyptian and American authorities focused their attention on Egyptians as they investigated the New York extremists. That focus appeared justified in that the Egyptians were more significant than others, such as the Palestinians, who became involved in the World Trade Center bombing. But even Egyptian nationals were peripheral to the Trade Center bombing. Mahmud Abu Halima was the only Egyptian directly involved, and it is unclear at what point he learned about the plans to build a bomb and when his participation in the conspiracy began. The emphasis on Egyptians caused one investigator to remark later, "We may have been focusing too narrowly."[5]

Any chance of catching the World Trade Center bomb in the making was lost as supervisors in Washington took a more sanguine view of developments in New York than did local investigators. In early 1992, while still working with Emad Salem, the regional FBI field offices in New York and New Jersey conducted a "preliminary inquiry" of the New York area extremists.

Such an inquiry can last for four months, with a possible three-month extension. But such an inquiry does not allow for wire taps or "mail covers," the scrutiny of outside envelopes for return addresses and postmarks. The request of the regional FBI offices to conduct a "full field" investigation, which would have allowed for both, was turned down by FBI headquarters in Washington, on the grounds that there was "no probable cause."

But investigators also missed the World Trade Center bombing conspiracy because it was a compartmentalized operation. The four defendants and two fugitives lived in five different places. Yousef and Salameh shared an apartment in Jersey City. Abdul Rahman Yasin lived with his brother in another Jersey City apartment. Abu Halima lived in Avenel, New Jersey. Ayyad lived with Arab roommates in Bloomfield, New Jersey, until taking his own apartment in Maplewood, New Jersey, after getting married in December. Ajaj remained in prison throughout this time. A careful analysis of the defendants' phone records reveals:

• Calls between the Yousef/Salameh and Yasin apartments were local calls and do not appear on the telephone bills. One cannot know the frequency of their communications.

• The Yousef/Salameh apartment called Nidal Ayyad at work throughout this time, but did not call him at home until he had his own apartment. Ayyad called them.

• The Yousef/Salameh apartment called Mahmud Abu Halima, and Abu Halima called them.

• Abu Halima and Ayyad never called each other. No evidence was presented that they knew each other well, and they were involved in different aspects of the bombing. Abu Halima participated in the activities at Pamrapo Avenue, where the bomb was made. Ayyad helped purchase chemicals and helped Salameh logistically, renting vehicles, scouting out the World Trade Center garage, and so forth, but he did not go to Pamrapo Avenue.

• Calls from the Yasin apartment to Ayyad began only on January 19, little more than a month before the bombing. Similarly, calls from the Yasin apartment to Abu Halima began only on January 21.

- There is no evidence that Ahmad Ajaj knew the conspirators, save for Yousef. But Ajaj did not call Yousef directly. When he spoke with him, it was by way of a friend in Dallas, Texas, who had three-way calling.

Thus, the picture that emerges is not one in which every suspect knew all the other suspects equally well and contacted them with equal frequency. Rather, Yousef and Salameh were at the center of the conspiracy, and the others had only limited contact with one another. Only in the conspiracy's last month did all the suspects come together. That made the conspiracy difficult to penetrate, and it is one more reason why the FBI, while investigating one of the conspirators, missed the conspiracy. It is also testimony to the care the conspirators took. It was a sophisticated operation.

Indeed, it is perhaps noteworthy that all the convicted conspirators (save for Ajaj, in prison) changed their phone numbers as the bombing approached. Nidal Ayyad got a new phone number when he moved into a new apartment on December 1, as he got married. Yousef and Salameh got a new number when they moved into the Pamrapo Avenue apartment on January 4. The number of the phone in the Yasin apartment was changed sometime between December 13 and January 4. And Abu Halima's number was changed on January 23.

Leaving the Extremists to Get Caught

Given the care exercised by the conspirators to avoid detection in the period before the February 26, 1993, bombing, it should be asked whether certain clumsy aspects of the bombing conspiracy were not in fact intended to set police on the trail of those left behind, the New York-based Muslim extremists. One investigator commented early on about those first arrested—Muhammad Salameh, detained on March 4, and Nidal Ayyad, arrested six days later. He said, "Their ineptness in evading detection suggested a conspiracy in which others masterminded the attack."[6] Indeed, Jim Fox of the New York office of the FBI told me that he had not wanted to arrest Salameh as early as he did.[7] Fox had wanted to let Salameh "walk around," to see whom he was dealing with. But Salameh's name was leaked to the media and he had to be arrested before it became public knowledge that authorities suspected him.

Two points are particularly suggestive. One involves the defendants' telephone bills. The other concerns the flight of only certain members of the plot, along with the failure of others to flee successfully.

The Telephone Bills. The telephone service of Ramzi Yousef and Muhammad Salameh was cut off for nonpayment in late January, a month before the World Trade Center bombing. They owed their long-distance carrier, Metromedia Communications Corporation, over $4,000. But they also had a Sprint Fone Card. They owed Sprint only $134. Additionally, Sprint asked for a $150 deposit, because the two did a considerable amount of long-distance calling and Sprint had no phone history for "Kamel Amer," the name Yousef gave the telephone company for the Pamrapo Avenue apartment.

Yousef did not pay the $284 that Sprint was asking for, although doing so would have allowed him to continue to make long-distance calls. Almost certainly, Yousef had that small sum of money. Why didn't he pay it?

After Yousef and Salameh lost their long-distance service, calls started to appear on the telephone bills of the other conspirators, linking them to the bombing. Nidal Ayyad began calling chemical companies to purchase materials for the bomb only after Yousef's long-distance service ended, presumably at Yousef's request.

Abu Halima applied for and received a Bell Atlantic telephone card at the end of January. Yousef and Salameh used Abu Halima's calling card between January 31 and February 12, when Abu Halima canceled it. Their calls left many incriminating numbers on his bill. Abu Halima received his telephone bill on February 26, the day of the bombing. He immediately told Bell Atlantic that a friend had used the card without his permission, and he asked that the calls be removed from his bill. They were, but to no avail. The request itself would appear as one more piece of evidence against Abu Halima.

It may well be that it suited Yousef to lose his telephone service, and then ask others for help, so that there would later be clear evidence linking Abu Halima and Ayyad to the bombing.

Those Who Did Not Flee. The flight, or nonflight, of the various World Trade Center suspects also suggests that the New York extremists were meant to be caught. Two suspects did not even try to flee, four fled, and one tried but failed.

Nidal Ayyad and Ahmad Ajaj remained behind. Both men, for very different reasons, thought they had nothing to fear. Ayyad was brazen. He was not involved directly in the actual construction of the bomb at Pamrapo Avenue, and evidently he thought he could not be held criminally responsible, although he had participated in other aspects of the bombing.

In the days following the bombing, Ayyad telephoned three New York newspapers and sent them letters claiming credit for the bombing in the name of the "Liberation Army, Fifth Battalion," with "Staff Lieutenant General [*al-Fareek al-Rokn*] Abu Bakr al-Makee" typed below. Ayyad's voice was subsequently identified on a recording left by an anonymous caller on the "News Tips" line of one newspaper. DNA testing of the saliva on an envelope containing one of the letters claiming responsibility for the blast revealed that the saliva matched his own.

Ayyad borrowed a radio from a colleague at work the day of the bombing, and then listened to an all-news station. The next day, Ayyad explained to the same colleague how a nitrate bomb could be made. Both actions were later cited as evidence against Ayyad during the trial. That behavior in the days after the bombing highlights the naïveté of the Palestinians involved. Ayyad was arrested on March 10.

Ahmad Ajaj—Wrongfully Convicted?

Ahmad Ajaj acted with a similar sang-froid. But it seems that he did so because in fact he had nothing to do with the bombing.[8] In jail at the time of the bombing, Ajaj was released on March 1, after having served six months for illegal entry.

His release, however, was a mistake. Ajaj was still supposed to be held, subject to a pending immigration deportation order. Unaware that any error had been made, Ajaj tried to find a permanent place to stay, as his uncle in Houston would not take him back. He asked for help from his designated parole officer, George Gonzalez, who tried to arrange temporary shelter for him in a Brooklyn mosque. The mosque, however, was reluctant. It put Ajaj up for one night, after which he slept in the subways.

Five days later, Gonzalez located a distant relative of Ajaj's in Dallas who agreed to take Ajaj in. As Gonzalez later explained, "I was curious why INS released him, so I called INS and was told he was released by mistake."[9] Ajaj was rearrested on March 9 in Gonzalez's office, to be held for possible deportation.

Investigators did not begin to link him with the World Trade Center bombing until April. He was charged with involvement in the bombing in mid-May, after authorities learned that he had entered the United States with Ramzi Yousef on a false passport.

Like the other defendants, Ajaj was convicted and sentenced to 240 years in prison. But it is not clear that Ajaj was even aware of the bombing plot, or that he understood that his traveling companion from Pakistan to New York, Ramzi Yousef, had masterminded the bombing. After all, Ajaj's role presumably was to create a diversion to facilitate Yousef's entry into America. Particularly as they had only known each other for a brief period, Yousef would have remained silent about his plans for the bombing, at least until they had both passed through immigration at JFK Airport. Until they had done so, Yousef would have had to assume that Ajaj might well face a prolonged engagement with U.S. law enforcement officials, as in fact he did, and that if he knew anything, he might talk.

Nor does it appear that Yousef could have revealed his plans subsequently to Ajaj. Yousef did not tell Ajaj about them over the telephone, as all Ajaj's phone conversations were taped and they contained no references to making bombs or to the World Trade Center.[10] Ajaj received no visitors and no mail while he was in prison. How could Ajaj have learned that Yousef was building a bomb in Jersey City? And even after the World Trade Center bombing, would Ajaj have known that Yousef had been responsible for it?

At the sentencing, Ajaj was the only defendant to assert his innocence, claiming, "Up to this very moment I do not know where the World Trade Center is. . . . I did not know anything of the bomb." Ajaj condemned the bombing as a "horrible crime," even as he harangued the courtroom with a long diatribe about Palestinian suffering at the hands of "fascist Zionist gangs" backed by the United States.[11] After two-and-a-half hours, Judge Duffy finally cut him off.

Israeli and Jordanian authorities have identified Ajaj as a petty counterfeiter. But the prosecution presented no real evidence during the trial to show that Ajaj was aware that a bomb was being built in Jersey City while he served out his prison term.

The Evidence. A focus of the government's charges against Ajaj is a December 29, 1992, telephone conversation between Yousef and Ajaj. The government claimed that Ajaj tried to arrange to send Yousef the materials—including the bombmaking manuals—he had carried into the United States, which had been seized by immigration authorities at the time of his arrest.

But there is little evidence for that. Ajaj wanted to ensure the safety of his belongings, but he did not want to send them to Yousef. Shortly before, at his December 22 sentencing hearing, the judge had ordered that Ajaj's personal effects be returned to him. Ajaj was concerned that they might be sent to a Texas address to which he did not want them to go. In their December 29 conversation, Ajaj suggested to Yousef that he would ask his attorney to have them sent to a different address. Yousef proposed, "Is it possible that I receive them?" an idea he pressed repeatedly.

Ajaj was reluctant. He first suggested that it would be better for Yousef to send someone else to pick them up. Later in the conversation, when Yousef proposed precisely that, Ajaj replied, "We'll see." Yousef raised the subject a third and last time, saying, "I will call your friend [the lawyer], so that I leave him an address. The address of someone who would come and pick up your belongings." Ajaj responded, "God willing"—that is, *Insha' Allah.*[12]

As used by Arabic speakers, that phrase can mean many things—fervent assent, indifference, or polite disagreement. Anyone dealing with an Arab bureaucrat who is told *Insha' Allah* is not pleased. By "God willing," the speaker means that the matter is not in his hands, so please do not expect an answer from him, at least not any time soon.

In fact, Ajaj did not even try to send his belongings to Yousef. A week later, on January 4, 1993, Ajaj's lawyer wrote to the court that his client had asked that his property be sent to the Houston address of Imran Mirza, Ajaj's immigration lawyer during his 1991–1992 asylum appeal. (See illustration 8–1; the letter reading "January 4, 1992," should read "January 4, 1993"; it reflects a New Year's clerical error.) However, that was not done. The material, for some reason, was held for "safekeeping," and it remained with the government.

It is clear that Ajaj did not want to send his belongings to Yousef. The bombmaking manuals never reached Yousef, and they were thus irrelevant to the bombing.

Ajaj himself did not seem to have been particularly interested in the bombing manuals. They were nothing special. The information they contained was legally and readily available in the United States. During the World Trade Center trial, a private investigator hired by the defense explained that he had readily checked out *The Anarchist Cookbook*, replete with bomb recipes, from the Mount Vernon Public Library, and that he had just as easily ordered two videotapes, "Homemade C4: A Closer Look" and "Deadly Explosives: How and Why They Work," from advertisements in a *Soldier of Fortune* magazine.

Ajaj knew all that, generally. Shortly after his September 1, 1992, arrest, Ajaj told friends that he did not expect a long sentence, despite the discovery of bombmaking manuals in his possession, because "similar ones are available in America."[13] Besides, Ajaj's manuals were about operations much simpler than the World Trade Center bombing. None of the bomb formulas in Ajaj's manuals quite matched that for the World Trade Center bomb, which was very rare. According to the government, urea nitrate had never been used as an explosive in a bomb abroad, and it had only been used once before in the United States, in a pipe bomb some five years previously.[14]

There were other items among Ajaj's belongings of much greater value than the bombmaking manuals. Ajaj's telephone conversations suggest that he was worried about two things. One was an airline ticket; and the other, his principal concern, was purloined identity documents. Ajaj brought into the United States twenty-two documents in the names of ten other people.

ILLUSTRATION 8–1
LETTER FROM AHMAD AJAJ'S LAWYER

 THE LEGAL AID SOCIETY
CRIMINAL DEFENSE DIVISION • FEDERAL DEFENDER SERVICES UNIT
EASTERN DISTRICT OF NEW YORK
225 Cadman Plaza East, Brooklyn, New York 11201
Tel.: (718) 330-1200   Fax: (718) 855-0760

Executive Director and
Attorney-in-Chief
Archibald R. Murray

Chief of Operations
Leonard F. Joy

Attorney-in-Charge
Thomas J. Concannon

January 4, 1992

Mr. Eric Bernstein, Esq.
Assistant United States Attorney
United States Attorney's Office
Eastern District of New York
225 Cadman Plaza East
Brooklyn, N.Y. 11201

<u>U.S.A. v. Ahmad Ajaj aka Khurram Khan, 92 CR 993 (RR)</u>

Dear Mr. Bernstein:

This letter confirms that my client Mr. Ahmad Ajaj has asked me to ask the government to send his property to Mr. Imran Mirza, Esq., 190 West Loop South, Suite 950, Houston, Texas 77027.

Please send me a confirmatory letter as soon as his property has been sent.

Thank you for your attention to this matter.

Sincerely,

Douglas G. Morris

Associate Attorney
(718) 330-1209

cc: The Honorable Reena Raggi
U.S. District Judge

Mr. Robert Heinemann, Esq.
Clerk of the Court

Defendant

SOURCE: Court files.

The jewels among them were three passports, all apparently stolen in Pakistan.

ABC News located the owner of one of those passports, Azan Muhammad, a young Pakistani with British citizenship, living in England. Azan Muhammad explained that his passport had been lost during a 1991 trip to Pakistan. Immigration stamps on the other two passports—one Saudi and the other Swedish, belonging to Khurram Khan (the passport that Ajaj had used to enter the United States)—suggest that they, too, were lost in Pakistan, around the same time.

Ajaj also had a U.K. driver's license belonging to Azan Muhammad; an international certificate of vaccination in the name of Khurram Khan; a Moroccan secondary education certificate belonging to yet another person; a copy of a blank Iraqi passport; and still more documents in other names. Ajaj was a counterfeiter. By altering details on the documents, perhaps with a Xerox machine, Ajaj produced false papers to corroborate the stolen passports and to create fictitious identities.

A document found in Ajaj's possession illustrates the technique. Illustration 8–2 reproduces a medical report from a Kuwaiti hospital, dated April 28, 1990, describing the recovery of Azan Muhammad from an automobile accident in which he had broken his nose. Ajaj's picture is on the medical report, and it has been assumed that Ajaj was the person who had the accident in Kuwait.

Azan Muhammad's passport, as found in Ajaj's possession, had an April 9, 1990, entry stamp into Kuwait, numbered "124." But Muhammad said that he had never been to Kuwait before he lost his passport in 1991. The Kuwaiti stamp in Muhammad's passport is a forgery, intended to make the Muhammad passport corroborate the Kuwaiti medical report. There is no Kuwaiti visa or exit stamp to accompany the entry stamp, as there should be. Indeed, the same entry stamp, number 124, dated April 9, with the year obscured, appears on Ajaj's Jordanian passport.

Moreover, if the Kuwaiti medical report did not belong to Azan Muhammad, neither did it belong to Ajaj. In April 1990, Ajaj was under the supervision of Israeli police. He was required to report twice daily to them. His picture had been added to the original report about an unknown third person.

ILLUSTRATION 8–2
MEDICAL REPORT FOR "AZAN MOHAMMAD"

MINISTRY OF PUBLIC HEALTH

KUWAIT — ARABIAN GULF

وزارة الصحة العامة

الكويت ــ الخليج العربي

Reference ....

الرقم ....

Date 28/4/1990

التاريخ ....

MEDICAL REPORT

Name: Azan Mohammad
Age: 20 years.
File No: 38802

Azan 20 yrs male was admitted in Adan Hospital on 17/4/1990 with the complains of bleeding face & broken nose resulted from a car accident.
On examination he had marked DNS to left side almost touching to lat.wal of nose. Union between bony and cartilagenous septum and a part of perpendicular plate of ehtmoid & vomer were broken. Ears & throat examination did not revel any abnormality. Sinuses were unclear and full of blood.

Pt. underwent operation, bleeding points were cathotrised & stopped, broken parts of the nose were custred. Sinuses were cleared and small broken strip of carilage was removed from floor.
Silastic-sheet was put on left side for 10 days.
He made uneventful recovery postoperatively and was discharged on 26/4/1990.

Dr. A. Chauhan
E.N.T. DEPT.
AL - ADAN HOSPITAL

GOVERNMENT
EXHIBIT
2799-4
SS 93 CR 180

CABLES : HEALTH KUWAIT			
	Admin. Affairs	Financial Affairs	Medical Stores
P. O. Box No.	5	1519	22575
TELEX No.	22729	22291	22745

برقيا : صحة الكويت

الوزارة المالية المستوردعات ص·ب

٢٢٥٧٥ ١٥١٩ ٥

تلكس ٢٢٧٢٩ ٢٢٢٩١ ٢٢٧٤٥

٣٠٠ ــ ٢٧ ــ ٩

SOURCE: Government Exhibit 2799, *United States v. Muhammad Salameh et al.*

The trove of documents that U.S. authorities had seized from Ajaj upon his immigration inspection would not be easy to replace. Those documents, not the bombmaking manuals, seem to have been Ajaj's main concern. That appears from Ajaj's telephone conversations. Ajaj was close to another Palestinian, Muhammad Abu Khdeir, proprietor of a Dallas hamburger shop. Abu Khdeir was also involved in Islamic activities, and Ajaj had known him from his earlier stay in Texas. Ajaj called him frequently from prison. Ajaj's telephone conversations were monitored, and Ajaj spoke to Abu Khdeir elliptically. He referred to his jihad activities as "studying" and to Pakistan as "the university."

In his first call to Abu Khdeir, on September 12, 1992, Ajaj expressed his concern about losing the materials that he had carried into the United States: "They've taken all my papers . . . all my telephone numbers . . . all the cards. . . . Everything!"[15]

Subsequently, Abu Khdeir tried to get that material from prison authorities for Ajaj. "Did you ask about the papers?" Ajaj asked Abu Khdeir on October 7. "I called three to four times, but nothing was there," Abu Khdeir replied. "I think they got stolen," Ajaj concluded. Ajaj also thought of an airplane ticket among his belongings. "Well, the ticket is lost now. May the Lord's vengeance be upon all nonbelievers."[16]

Similarly, in his December 29 discussion with Yousef, Ajaj was concerned, above all, with the passports and identity papers. "I don't want my passport to get lost . . . and the original university diplomas and things. You know the university diplomas."[17] Of course, there were no university diplomas among Ajaj's possessions. He was referring to the identity documents.

Nor is it likely that Yousef himself was particularly interested in the bombmaking manuals, either, as he could easily obtain the same information elsewhere. Yousef, too, seems to have been most interested in the treasure trove of identity papers. Indeed, the irrelevance of those "bombmaking manuals" became clear, following Yousef's aborted plot to bomb twelve U.S. airplanes. The books and handwritten instructions he used were far more sophisticated than anything Ajaj ever had in his possession.

The transcripts of Ajaj's prison conversations—from September 12, 1992, the first presented into evidence, until April 12,

1993, the last—provide a consistent, coherent picture. Ajaj was strongly, even fanatically, committed to the Palestinian and Islamic cause. His role was to forge identity documents for his compatriots, and he looked forward to getting out of prison and resuming that activity. Although there is no reason to question the other verdicts, Ajaj seems to have been the victim of prosecutorial excess.

Those Who Fled or Tried to Flee

After the bombing, four people fled: the mystery man, about whom practically nothing was known; Ramzi Yousef; Mahmud Abu Halima; and Abdul Rahman Yasin. Also, Muhammad Salameh tried to flee.

Regarding the mystery man, investigators did not know either his nationality or the name under which he had entered or left the United States. "Rashid" was the name that Yousef used in New York, and the mystery man was known to the others only as "Rashid's friend." As Ayyad later explained to authorities, Salameh told him before the bombing,

> Rashid's friend is coming to help us. He arrived one week prior to the bombing. . . . The night of the bombing, Salameh took Rashid and his friend to John F. Kennedy International Airport for flights out of the country.[18]

As we have seen, Yousef's very elaborate preparations for escape had already begun in December, with his phone calls to Baluchistan and his application for a new passport in the name of Abdul Basit Mahmood Abdul Karim. On February 23, three days before the bombing, he purchased a ticket for Pakistan—the only known conspirator (along with his "friend") to buy his ticket before the bombing.

The evening of the blast, Abu Halima appeared at a car service company where he occasionally worked. As the proprietor testified, Abu Halima "looked very scared, very nervous." Abu Halima told him that there had been an "accident" and he began to pray in the office, something he had never done before. The next day Abu Halima purchased a ticket for a March 1 flight to Saudi Arabia, from where he went on to Egypt. He also bought tickets to Cairo, for his wife and four children to follow him. Abu Halima paid cash for the lot, later telling fellow inmate Theodore Williams that the money for the tickets had come from abroad.

But despite the money he had, Abu Halima's flight was not well considered. He had no secure place to go. He went to his family home in Egypt and remained there, despite the arrests of Salameh and Ayyad. Egyptian authorities arrested Abu Halima on March 16.

Abdul Rahman Yasin bought a ticket to Jordan on March 1 for a flight departing March 5—Royal Jordanian Flight 262. He took that plane the day after he was picked up, questioned, and released by New Jersey FBI. From Jordan, Yasin went on to Baghdad.

Flight 262 stops in Amsterdam. Muhammad Salameh also bought a ticket on March 1 for Flight 262, departing March 5. But Salameh was short of money. He did not have enough for an adult ticket to Jordan, let alone Europe. He bought an infant ticket only as far as Amsterdam for $65.45, which he used to get a visa from the Dutch consulate.

Salameh was arrested on March 4 as he returned for the fourth time to the rental agency in an effort to recover his $400 deposit on the van. Salameh has been ridiculed for that, and 97 percent of Americans say they would not have done as he did.[19] But Salameh undoubtedly needed the money to buy an adult ticket, so he could actually get on the plane. If those most responsible for the bombing had wanted to ensure that Salameh did not get caught, they surely could have provided him the few hundred dollars he needed. More likely, those most responsible for the bombing directed the doltish Salameh back to the rental agency specifically to facilitate his arrest.

The Conceptualization of the World Trade Center Bombing

If the conspirators had indeed succeeded in causing the death and destruction they planned, the bombing would have been a historically traumatic event on the scale of Pearl Harbor or the sinking of the *Maine*, both of which sent America to war. On such dramatic occasions people want immediate answers. There would have been immense pressure on authorities to produce them.

Imagine, further, that following such a catastrophic attack, émigré extremists like Salameh and Ayyad were soon arrested. Although Islamic extremists have ties with several Muslim governments, many Americans associate them with Iran, in part a legacy of the Iranian hostage crisis and the civil war in Lebanon, when Iranian-backed extremists seized Americans there. Would the new and inexperienced Clinton administration, just one month in office, have jumped to the hasty conclusion that Iran was behind the Trade Center bomb?[20] If so, what would it have done? Serious military action against Iran would have been one possibility.

But would the argument also have been made that UN sanctions should be lifted on Iraq because Iraq, in its weakened state—after its devastating defeat in the Gulf War—posed little danger, and because Iraq could help the United States deal with the threat from Iran, apparently striking America's shores with such extraordinary viciousness? The Trade Center bombing may well have been, simultaneously, Saddam's revenge *and* an attempt to promote the lifting of sanctions.

Indeed, a decade earlier, Iraq had carried out another act of terrorism, similar in that Baghdad had had a strategic aim it hoped to achieve by setting up others to take the blame. Iraq was behind Abu Nidal's attempted assassination of Israel's ambassador to Great Britain, Shlomo Argov, in June 1982, as discussed in chapter 2.[21]

At that time Menachem Begin, Israel's prime minister, and Ariel Sharon, the defense minister, wanted to attack the PLO in Lebanon. But they were inhibited from doing so because the United States had during the previous year brokered a cease-fire between Israel and the PLO. The attempted assassination of Argov provided the pretext Begin and Sharon needed to break the cease-fire with the PLO. Although Israeli intelligence immediately reported that Abu Nidal, whose organization was at bitter odds with the PLO, was responsible for the assassination attempt and that the PLO was not involved, Begin paid little attention. On June 6, Israel invaded Lebanon.

The Iran-Iraq war had by then reached a critical juncture.

An Iraqi collapse was widely expected. Israel's assault on the PLO served Iraq's purposes by tying up Syria, which had been providing military support to Iran. Iraq also tried, unsuccessfully, to use the Israeli invasion to embarrass Iran into ending the war, stressing the urgency of an "Islamic reconciliation" to defend against "the forces of Zionism."

Of course, the shooting of Argov did no more to advance Iraq's strategic goals than did the World Trade Center bombing. But it reflected the same mind-set. A terrorist act was committed on the basis of assumptions about how the victim would misjudge the party responsible. Had the Trade Center tower fallen and large-scale casualties ensued, contemporary attitudes toward Iran and Islamic extremism could have provided fertile soil for Iraq to manipulate U.S. opinion toward the hasty conclusion that Iran was behind the bomb and that it was time to ease the postwar constraints on Iraq. Or that, at least, may have been Saddam's calculation.

9

The Question of State Sponsorship

Indeed, the simplest, most straightforward explanation of the World Trade Center bombing is that it was an Iraqi intelligence operation, with the Muslim extremists as dupes. That, in fact, is what New York FBI suspected. Shortly after the bombing, an FBI agent, John Anticev, speculated to Emad Salem,

> Do you ever think that Iraqi intelligence might have known of these people who were willing to do something crazy, and that Iraqi intelligence found them out and encouraged them to do this as a retaliation for the bombing of Iraq. . . ? So the people who are left holding the bag here in America are Egyptian . . . or Palestinian. . . ?
>
> But the other people we are looking for, Abdul Rahman, he is gone. . . . I hate to think what's going to happen if this guy turns out to be, if this turns out to be an Iraqi intelligence operation . . . and these people were used. . . .
>
> If he's out of the country, then that is a very big indication that he is directly involved in this and he, and he split.[1]

Speaking with another FBI agent, Salem remarked, "Is lot of suspicious about Iraqi intelligence involve." She replied, "Yeah."[2]

In June 1993 I met with Barbara Jones, the number two person at the New York district attorney's office. She also said there were suspicions that Iraq was behind the bomb. Indeed, in the course of this work, I came to understand that the majority of New York law enforcement—including the New York FBI, as well as other law enforcement agencies—suspected that Iraq was behind the Trade Center bombing.

Regional FBI offices, however, must adhere to decisions and policies set in Washington. And the notion that Iraq seemed to be behind the Trade Center bomb was evidently not a line of inquiry Washington was interested in pursuing. Rather, once initial arrests were made, the Justice Department took the position that it had cracked the case and caught the perpetrators, or most of them, and was bringing them to justice.

A similar problem existed years ago regarding the Mafia. For decades, J. Edgar Hoover refused to admit that America had an organized crime problem, until Bobby Kennedy, as attorney general, forced him to do so. It is easier to go after the small fry than to catch the big fish, and law enforcement is always vulnerable to the temptation to cut off a conspiracy at the most convenient point rather than seek its head.

Thus, six weeks after the World Trade Center bombing, five Arabs were under arrest. A sixth man—the mastermind, Ramzi Yousef—had fled. At that point, in early April, the Justice Department announced that it had captured most of those involved. The bombing, it asserted, was the work of a loose group of Muslim extremists with no ties to any state.

The predictable media frenzy followed, vividly depicting the loose network of religious extremists said to be behind the Trade Center bombing. But the obvious question was not asked: How could the government know that those it had arrested had no ties to any state? It was still early in the investigation, and the government's understanding of the bombing was incomplete. Two of those arrested, Ibrahim El Gabrowny and Bilal al-Qaysi, never stood trial for the bombing, and two others, Abdul Rahman Yasin and Ahmad Ajaj, would be indicted for it only later, even as the mystery man remained unexplained.

Thus, when the government announced that the bombing had no state sponsorship, it had not even determined definitively who was involved. And even if it had, it could not yet have con-

ducted an investigation of sufficient thoroughness to conclude
that the bombing had no state sponsorship. Such an investigation would have entailed, at a minimum, a meticulous examination of all records associated with the defendants, to ensure that
they had had no contact with foreign intelligence agencies, or at
least none that could be found. Such a time-consuming process
could not have been accomplished in six weeks.

Moreover, the mastermind, Ramzi Yousef, was a fugitive,
about whom little was known. How could anyone be sure that he
was not a foreign intelligence agent, even an Iraqi agent? After
all, he had come to the United States on an Iraqi passport and
had been known in New York as "Rashid, the Iraqi." Indeed, the
most savvy of the conspirators among the New York extremists,
Mahmud Abu Halima, came to understand, either on his own
or from his Egyptian interrogators, what had happened. Abdul
Rahman Haggag, an Egyptian extremist arrested in the second
New York City bombing conspiracy case, cooperated with the government. Haggag explained to authorities that Abu Halima had
told him,

> Ramzi Yousef showed up on the scene and brought a
> number of individuals together and escalated the initial
> plot. Mahmud stated that Ramzi Yousef used himself
> and others as pawns and then immediately after the
> blast left the country. Mahmud advised that the initial
> plan was not as large as what happened and that Ramzi
> Yousef was responsible for bringing the individuals re-
> sponsible together and making it happen. Mahmud also
> made reference that law enforcement authorities would
> not be able to catch Yousef. [Haggag] questioned why,
> and Mahmud responded, "Don't ask."[3]

As Abu Halima's younger brother similarly explained, "Muhammad Salameh had dealt with Iraqi intelligence."[4]

Yet that is not what less sophisticated individuals involved
in the bombing, like Salameh and Ayyad, understood. Ayyad told
authorities how the plot emerged, or at least how he understood
it. He, Salameh, and Bilal al-Qaysi, the three Palestinians
among the five New York–area extremists first arrested, had long
wanted to build bombs. In the spring of 1991, they had even purchased small amounts of chemicals from a Syrian in Brooklyn.

Although they had some manuals, they could not figure out what to do with the chemicals they had purchased and "realized they needed someone who was an expert at making bombs to help them." They stored the material until the autumn of 1992, when Salameh told Ayyad, "I have this guy from Afghanistan willing to do anything."[5] And that is the ultimate origin of America's understanding of the World Trade Center bombing. It is the understanding of Muhammad Salameh. It is difficult to overstate Salameh's naïveté. Yousef had told him, "I will not forget about you. If you are arrested, I will do something, kidnap or whatever and demand your release." So, Salameh asked Ayyad, "Do you want your release demanded, too?"[6]

And Washington, evidently, was more inclined to accept that view than to challenge it. The bombing occurred just one month after the inauguration of a new president, and at a time of major transition within the Justice Department. The FBI director, William Sessions, was fighting a losing battle for his career. He would be removed in July and replaced by Louis Freeh. Also, the Clinton administration had problems choosing an attorney general. One was not sworn in until mid-March. Meanwhile, the number two person at the Justice Department was in charge—Webster Hubbell, a close associate of the Clintons who was later forced to resign and convicted of fraud.

Iran or Iraq? The most tangible signs of foreign support for the World Trade Center bombing were the two known fugitives, Ramzi Yousef and Abdul Rahman Yasin. In fact, during the ABC/*Newsweek* investigation, we focused on them, assuming them to be the key to any state sponsorship. But whom could they have worked for? The usual suspects are Iran, Iraq, Libya, Syria, and Sudan. There were no indications that any of the last three were involved in the World Trade Center bombing, and no one has seriously suggested they were. That leaves only two possibilities: Iran and Iraq.

No evidence ever emerged linking Iran to the bombing. Iran reportedly gave money to Shaykh Omar, according to intelligence sources. But even if those reports are true, they do not link Iran with the Trade Center bombing. Shaykh Omar was not the operational director of the terror. He instigated and inflamed, but he did not order specific bombings. Nor did any evidence ever emerge that the World Trade Center bombers received money from him. Indeed, if they had, Salameh would probably have had the money to buy the airline ticket he needed to flee.

Moreover, despite the conspirators' many overseas telephone calls, there were none to Tehran or to any area in Iran, save Baluchistan. Thus, the telephone records do not suggest Iranian sponsorship.

Iran's established modus operandi is to use professionally trained terrorists for major bombings, relying on local supporters only for secondary aspects of the operation. The 1992 bombing of Israel's embassy in Buenos Aires is an example of Iran's style. And if Iran was behind the 1994 bombing of Buenos Aires's Jewish community center, as is generally assumed, the same principle applies. No one was ever caught. Those who carried out the bombing came and went. By contrast, in the World Trade Center bombing, amateurs played central roles and, in fact, seem to have been deliberately left behind to get caught.

Would Iran have run an operation in which Muslim extremists were set up to be caught? And why should Iran attack the United States in February 1993? During the years of the most intense American-Iranian hostility, in the late 1970s and early 1980s, Tehran struck American targets abroad. Why would Iran escalate the confrontation much later by bringing it onto American soil?

Moreover, Iran has always used Shi'i, whether Arabs or Iranians, to carry out major attacks. There are no Shi'i associated with the World Trade Center bombing.[7] All of those charged and convicted were Sunni. They had associations with Sunni countries—above all, with Saudi Arabia. Ibrahim El Gabrowny went to Saudi Arabia to raise money for Nosair's defense, and Shaykh Omar spent three years, from 1977 to 1980, in Saudi Arabia.

Indeed, following Ramzi Yousef's arrest, it became clear that he was a Sunni Muslim who was strongly anti-Shi'a, or at least professed to be. The South African Muslim who turned him in, Ishtiyaak Parker, explained that Yousef had appealed to him on sectarian grounds; that is, Sunni versus Shi'a.

Americans may be unfamiliar with the nature of the Sunni-Shi'a divide. Formally, it is a dispute about the eighth-century succession to the role of the caliph, successor to the Prophet Muhammad and leader of the Muslim community. In practice, it is

a dispute about the character of Islamic authority and the interpretation of its laws and texts. Sunni Islam tends to be more
authoritarian, more in the service of rulers. Shi'a Islam allows
for a more independent clergy.

Such disputes can be deeply felt where Sunni and Shi'i rub
elbows, and Pakistan is one area where they do. Pakistan is over-
whelmingly Muslim, but 20 percent of the population is Shi'i.
Saudi Arabia, the most orthodox Sunni state, and Iran, the most
orthodox Shi'a state, both fund Sunni and Shi'i extremist groups
in Pakistan, where their intense rivalry contributes to that country's endemic violence. It is almost inconceivable that Ramzi
Yousef, a Sunni Muslim who appealed to anti-Shi'a sentiments
in his dealings with others in Pakistan, could have been working
for Shi'i Iran. Indeed, there is almost nothing to suggest an Iranian role in the World Trade Center bombing.

The Fugitive in Iraq

With Ramzi Yousef's arrest in February 1995, one indicted fugitive was left in the World Trade Center bombing: Abdul Rahman
Yasin. He is charged with having mixed chemicals for the bomb.
That is a more central role than the one played by Nidal Ayyad,
who kept his distance from the bombmaking site, or Ahmad Ajaj,
who was in jail while the bomb was made and, as we have suggested, is probably innocent.

That Abdul Rahman is in Baghdad is no secret. Following
Abdul Rahman's March 5 flight, as suspicions about his role
grew, New Jersey FBI regularly called his brother, Musab Yasin,
into its office. And at the FBI's request, Musab regularly called
Abdul Rahman in Baghdad, asking him to return to the United
States to answer a few questions for the FBI.[8] Abdul Rahman
would say that he was going to do that. Only, he just had a few
more things to finish up. As Jim Fox, former head of the New
York FBI office, remarked, even after Abdul Rahman had successfully fled the country, he continued to fool some of the FBI
agents involved.[9] Abdul Rahman was finally indicted on August
4, 1993. Perhaps notably, that was a decision made in New York.
Washington was merely informed about it, as Gil Childers, the
lead prosecutor, explained.[10]

And given the totalitarian nature of the Iraqi regime, Musab's phone calls to his brother take on special significance. A regime like Iraq's, by its nature, seeks absolute control and is suspicious of that which it does not control. Moreover, every phone in Iraq is tapped, or at least potentially so. If Abdul Rahman had participated in a major bombing conspiracy, like the Trade Center bombing, and he was *not* working for the Iraqi government, he could be in very serious trouble. The Iraqi government would be extremely suspicious of an individual whom it did not control, who had been involved in a major act of terrorism. After all, Abdul Rahman could become involved in bombings in Baghdad. Thus, Iraqi intelligence would almost certainly pick him up for a very harsh interrogation, if not do away with him entirely.

And Musab would be aware of that. Musab is a U.S. citizen. Nothing that U.S. authorities could have done to him could remotely resemble what Iraqi intelligence could do to his brother. Therefore, it is highly unlikely that Musab would have called his brother in Baghdad from the FBI's office, advising him that the FBI wanted to speak to him, unless Musab believed that his brother was working for Iraqi intelligence.[11]

Indeed, a year later, in the spring of 1994, a Jordanian-based stringer working for ABC News spotted Abdul Rahman outside his father's house in Baghdad and learned from neighbors that Abdul Rahman worked for the Iraqi government. After ABC broadcast its story in June 1994, Iraqi authorities took Abdul Rahman and the other men in the house away to an unknown location. His sixty-five-year-old mother, ill with cancer, was allowed to visit them until she died in October 1994, in a hospital reportedly run by Iraqi security.[12] And as recently as May 1998, FBI director Louis Freeh affirmed that Abdul Rahman was in Iraq.[13]

The question of Abdul Rahman's presence in Baghdad is rendered even more significant by UN Security Council Resolution 687, the formal cease-fire to the Gulf War. In paragraph 32, Iraq committed itself to renouncing all forms of support for international terrorism. Minimally, Baghdad is harboring a terrorist— one form of support for international terrorism. According to Allan Gerson, a former counsel to the U.S. mission to the United Nations, paragraph 32 imposes the obligation under international law to report suspected acts of Iraqi support for international terrorism to the UN secretary general.[14] The United States has never done that. Nor has it made any real effort to secure Abdul Rahman's extradition.

That is so, although the United States has an extradition treaty with Iraq that dates from 1934. The Reagan administration used that treaty in 1985 to press for the extradition of Palestine Liberation Front chief Abu Abbas, for his role in the hijacking of the *Achille Lauro* and the murder of Leon Klinghoffer. Although it demurred on secondary issues, Baghdad accepted the treaty's basic legitimacy. Thus, there is recent diplomatic precedent for both American and Iraqi acknowledgment of the validity of the extradition treaty.

Moreover, Abdul Rahman had an American passport. That gave U.S. authorities even greater right to claim that as an American citizen, he should be brought to the United States and tried in an American court. Indeed, in the unrelated case of the murder of an American woman by her husband, a naturalized U.S. citizen who fled to his native Jordan, U.S. authorities pressed for the man's extradition, although the United States then had no extradition treaty with Jordan. According to the *New York Times*, the American request to the Jordanian government was based on "an international legal principle that the crime was committed on American soil, by an American defendant, on an American victim."[15] The same principle applied to Abdul Rahman Yasin. And his crime was infinitely more serious. Indeed, the Clinton administration's failure to do anything significant about the presence of an indicted Trade Center bomber in Baghdad borders on the inexplicable. It is one of several enigmas suggesting that the administration is not keen to draw attention to anything that points toward an Iraqi role in the attack.

During the 1991 Gulf War, Iraq repeatedly threatened revenge, promising that the Iraqi people will "avenge the pure blood that has been shed no matter how long it takes."[16] One commentary even warned "Bush and his accomplices" that Iraq would pursue them "even if they leave office and disappear into oblivion. . . . They will understand what we mean if they know what revenge means to the Arabs."[17] At the time, the threats seemed vain and empty. But they no longer seemed so after April

1993, when Kuwaiti authorities discovered a car bomb intended for George Bush and his entourage. And one of Iraq's wartime threats was, "The American arena will not be excluded from the operations and explosions of the Arab and Muslim mujahedin and all the honest strugglers in the world."[18]

The World Trade Center bombing occurred on the second anniversary of the Gulf War's end.[19] Arabs are notorious for celebrating such anniversaries. The State Department issues anniversary watch lists, and security is routinely stepped up at U.S. installations abroad. Shortly before the Trade Center bombing, Neil Gallagher, head of FBI counterterrorism during the Gulf War, remarked that the problem of Iraqi terrorism, which had preoccupied him two years before, had not ended. "There is a saying in the Middle East that the best revenge is served on a cold platter. . . . The Iraqis don't have time pressure. It could be done on an anniversary date."[20] Indeed, the FBI issues annual reports on terrorism. They are written in the year after the events they describe. The 1990 and 1991 reports both warned of the danger of Iraqi terrorism. As the 1991 report, written in 1992, observed, "The ultimate military defeat of Iraq certainly diminished the threat of direct Iraqi terrorist activity. However, the threat was not eliminated."[21]

Indeed, the UN weapons inspectors stumbled upon evidence of active Iraqi involvement in terrorism. In the summer of 1996, an UNSCOM weapons inspection team led by Scott Ritter sought to inspect a building. It turned out to be a school for terrorism. It was run by Directorate M-21 of Iraqi intelligence, the directorate for "special operations," which reported to Directorate M-4, responsible for covert operations. Because some employees had tried to sneak files out of the facility, the weapons inspectors felt obliged to examine the papers there. "Document after document outlined an international program of terror," Ritter explained. "These were state-sanctioned assassins, who did not shrink from shedding the blood of innocent civilians, including women and children. It was all justifiable to them in the name of defending the regime."[22]

On March 19, 1994, Tariq Aziz, Iraq's deputy prime minister, was asked about allegations of Iraqi involvement in the World Trade Center bombing. Aziz categorically denied the charges, claiming to know nothing about the fugitives, Ramzi

Yousef, or even Abdul Rahman Yasin. In fact, the essence of his reply was that Iraq knew nothing about the bombing.[23]

Yet less than a month later, Iraq learned of the ABC/*Newsweek* project to investigate state sponsorship in the Trade Center bombing. Baghdad offered to share information about the bombing. The offer was broadcast on April 12 on Baghdad's official media, which explained, "Iraqi intelligence services have acquired documented and serious information regarding the incident of the explosion at the World Trade Center in New York."[24] The Iraqi Interests Section in Washington—an office under the authority of Algeria that formally represents Baghdad in the absence of diplomatic ties between the United States and Iraq—immediately repeated the offer, as did Iraq's information minister. When John Wallach, a Hearst journalist, visited Baghdad in May, Wallach was stunned when both Aziz and Foreign Minister Muhammad al-Sahaf brought up the subject in separate interviews.[25] Wallach had come to Iraq to learn about the effects of sanctions; the World Trade Center bombing was far from his mind.

But in June, when the ABC/*Newsweek* investigation aired, they pulled their punches. They never pointed a clear finger at Iraq. Jim Fox had been a big help to the journalists, but Washington had "spun" them away from Iraq. And Baghdad never delivered any information on the bombing. The offer was conditional on the proviso that a "competent authority" from the U.S. government come with a "neutral party," a U.S. congressman or a French official, to Baghdad or Geneva. The State Department responded that it would take any information through regular channels—the Iraqi Interests Section in Washington. Baghdad did not reply.[26]

The early suspicions of New York law enforcement that Iraq was behind the bombing never changed over the course of the investigation or the trial. After the March 4, 1994, verdicts against the first set of Trade Center defendants, Jim Fox told the *Chicago Sun Times* that Yousef appeared to be guided by a foreign intelligence agency and that Iraq, seeking revenge for the Gulf War, might well have been behind the bombing.[27] Later, Fox told me that the "majority of senior law enforcement officers in New York believe that Iraq was involved."[28]

Egyptian officials also saw an Iraqi hand in the bombing. They questioned Abu Halima under torture, before extraditing him to the United States. He told them about the involvement of two Iraqis, who had managed to flee. Egyptian officials understood that Iraq was behind the bombing, according to an Egyptian intelligence source who spoke to *Newsweek's* Chris Dickey. Egyptian authorities mentioned the two Iraqis to U.S. officials, but they did not stress the point, in part because of the mistrust that then existed between Cairo and Washington on account of Shaykh Omar's presence in America. Still, the Egyptians assumed that the Americans would understand that Baghdad was behind the bombing. They were surprised when the Americans did not appear to grasp that insight.

Indeed, the Egyptian press reported at the time that Iraqi intelligence was behind the bombing.[29] But the Egyptians had their own interests. The Egyptian government, along with other Arab governments under assault from Muslim extremists, wanted to link the extremists with Iran, to discredit them as an alien element in Arab society (Iranians are Persian, not Arab, in ethnicity). Thus, Cairo had little to gain by playing up Iraq's suspected role in the Trade Center bombing. In fact, Egypt's purposes were best served when Shaykh Omar was arrested and came to be seen as the party most responsible for it.

If there is so much evidence to link Iraq with the World Trade Center bombing, why isn't it better known? That is a big question, which we will try to answer in this and subsequent chapters. Part of the answer is that the Trade Center bombing was treated primarily as a criminal matter. Attention focused on the trials of the individual defendants. The American public mistook the "guilty" verdicts for an explanation of the bombing and did not understand that the trials were *not* directed toward addressing the question of state sponsorship. Indeed, as Ramzi Yousef's trial for the Trade Center bombing began in August 1997, a former prosecutor who had participated in the first trial, Henry De Pippo, cautioned, "There is certainly a legitimate public curiosity to the question of, was there a greater force behind this, and if there was, who was it? . . . But you're not likely to get the answers to [these] questions in the context of a criminal trial because the prosecution is much more limited."[30]

Also, many people think of Baghdad as a secular regime and do not understand how seemingly antagonistic ideologies—Arab nationalism and Islamic extremism—can work together. Saddam and his cronies are irreligious, to be sure. But they have long been willing to exploit the extremists' strong passions for their own purposes.

In the mid-1970s, Baghdad was engaged in an intense rivalry with Damascus. It sought to cultivate ties with secular, nationalist groups in Syria. But it failed. And so it turned to Syrian extremists. In the early 1980s, in cooperation with Jordan, Iraq sponsored a bloody Islamic terrorist campaign. In the most notorious incident, in November 1981 in Damascus's Izbekia gardens, nearly 200 people were killed by a car bomb. The belief that Saddam's regime could not or would not work with Islamic extremists is mistaken.

There is also a widespread feeling that Saddam would not dare to do something like blowing up the World Trade Center towers. America taught him a lesson during the Gulf War, didn't it? Iraq is suffering badly from economic sanctions. Saddam desperately wants them lifted. If Iraq were caught committing acts of terrorism against America, sanctions would never be lifted, and perhaps worse misfortunes would befall Iraq.

But that is American pragmatism. If Saddam saw his situation in those terms, would he have tried to assassinate George Bush in April 1993? And if Saddam believed that cooperating with the United States and appearing to comply with the UN resolutions were the best way to get sanctions lifted, would he have lunged at Kuwait in October 1994? Would he have engineered the series of crises from October 1997 through December 1998 that ended the presence of the UN weapons inspectors in Iraq?

A Kuwaiti professor of political science once advised that there was something very important Americans had to understand about the mentality of Saddam and those around him. In the late 1980s, he said, during the latter years of the Iran-Iraq war, he was a member of an Arab academic delegation that visited Iraq and met with Tariq Aziz. They asked Aziz why Iraq hit tankers carrying oil from Iran, even when the tankers belonged to countries that were friendly to Iraq. Aziz replied that Iraq wanted more international pressure to end the war with Iran, and "The way to get people to do what you want is to hurt them."[31]

Saddam has maintained an enduring contempt for American society. His contempt is matched by a deep-seated instinct to revert to threats, bullying, and violence to get his way. And, of course, there is also his thirst for revenge.

10

Bill Clinton, America, and Saddam Hussein

Saddam's August 2, 1990, invasion of Kuwait was premised on a deep disdain for the United States. He believed that America would not respond militarily, and that if it did, Iraq would prevail. As Saddam told April Glaspie, the U.S. ambassador in Baghdad, shortly before Iraq invaded Kuwait, "Yours is a society that cannot accept 10,000 dead in one battle."[1] The Gulf War did not really disabuse Saddam of that notion. It demonstrated only that America's high-tech weaponry could force Iraq to withdraw from Kuwait without America's suffering significant casualties.

Rolf Ekeus was the first chairman of the UN Special Commission, UNSCOM. UNSCOM was established in the spring of 1991 for the purpose of finding and destroying what remained of Iraq's chemical, biological, and long-range missiles programs after the war. Once UNSCOM reported that it had finished its job, the sanctions were to be lifted, according to UN Security Council Resolution [UNSCR] 687, the formal cease-fire to the Gulf War. But Saddam never gave up those weapons programs—a fact recognized only after his son-in-law, Hussein Kamil, defected in August 1995. Thus, as Ekeus advised the U.S. Congress in 1996, "The Iraqi government does not consider the

Gulf War was a war with an ending; the struggle is still going on. It was a battle of Kuwait, not a war of Kuwait."[2]

America was seriously divided before the Gulf War began. Many people wanted to rely on sanctions to force Iraq out of Kuwait rather than go to war, and afterward, many took the position that enough had been done to Saddam and his country. The only proper work remaining was to use the good offices of the United Nations to ensure that Iraq's weapons of mass destruction were fully destroyed. Anything more, like ousting Saddam, went beyond what America should do. It soon became unfashionable to "demonize" Saddam or "personalize" the conflict with Iraq. In that view, the American elite was at sharp odds with the American public, which came to believe that the United States should have continued fighting the Gulf War until Saddam was overthrown.[3]

Yet in the period after the war, President George Bush hesitated in dealing decisively with Saddam. The result was a game of cat and mouse—or "cheat and retreat," as Saddam's strategy was dubbed, and "threat and forget," the derisive motto coined for Bush's. Several times in 1991 and 1992 he geared up to strike Iraq, to force it to comply with UNSCOM's demands—only to back down as some ambiguous compromise emerged. Still, Bush proved far more willing to support UNSCOM's operations with American military muscle than did his successor.

That Saddam's opinion of America did not fundamentally change is perhaps understandable, given the dynamics of the postwar period. The Gulf War may have taught Saddam a certain measure of respect for George Bush, but it did not alter his underlying contempt for American society. Saddam's advisers would likely have told him of the uncertain position Governor Bill Clinton had adopted in late 1990 and early 1991, as America debated whether to go to war. And a consideration of the World Trade Center bombing "time line" suggests that the final decision to carry out the bombing was based, at least in part, on Saddam's disdain for the president-elect.

On the night of November 3, 1992, Americans learned that George Bush had lost the presidential election. It was already morning in Baghdad. That same day Saddam went to Ramadi, a Sunni Arab stronghold in western Iraq, and celebrated by firing his pistol into the air to show that "the Mother of Battles . . . has continued and will continue."[4] Little did America understand what he meant.

Not until November 1992 did the World Trade Center conspirators actually begin building their bomb. On November 18, the first of what would prove to be many calls to chemical companies appeared on Ramzi Yousef's telephone bills. At the same time, Yousef began calling surgical supply companies for equipment that included the gloves, masks, and rubber tubing they would use in making the bomb. On November 20, calls from a World Trade Center telephone booth began to appear on the Yousef/Salameh telephone bill. On November 30, Salameh rented a storage locker for the chemicals and Yousef began buying them.[5]

Supposing that a bombing plot had existed as early as the summer of 1992, into which Salameh was already recruited; and that Yousef was in New York as early as September, and living with Salameh by October; then why did Yousef wait until November to begin building the bomb?

It looks as if the final go-ahead for the operation was given only after Bush lost the election. A megalomaniacal, vengeful figure who rules by force and, in the end, respects only force, Saddam is almost a parody of himself, and is capable of making a parody of others.

Bush's Last Days in Office and Clinton's First Days

Saddam took immense satisfaction in Bush's electoral defeat. As Bush's term drew to an end, Saddam initiated a series of challenges intended to show that he, not Bush, had prevailed. On December 27, 1992, an Iraqi jet violated the "no fly zone" in southern Iraq, and U.S. planes shot it down. When Iraq subsequently redeployed SAM missiles in the area, the United States did not respond. Only on January 6, 1993, as yet more problems arose with Baghdad, did the United States, backed by France, Great Britain, and Russia, issue an ultimatum that Baghdad remove its missiles from the area. But Iraq did not do so.

On January 7 Baghdad began to block the travel of UN weapons inspectors into Iraq, placing conditions on their flight. On January 10, after a series of small-scale raids on Kuwaiti border posts, some 500 Iraqis crossed into Kuwait and seized a cache of weapons, including four Silkworm ship-to-shore missiles, carelessly left within reach of the Iraqi marauders by UNIKOM, the UN's Iraq-Kuwait border observer mission. On January 11 and 12, several hundred Iraqis again crossed the Kuwaiti border.

With much reluctance, Bush reacted, as American planes attacked anti-aircraft missile sites in southern Iraq on January 13. Baghdad claimed that nineteen people were killed and fifteen wounded, many of them civilians. What, if anything, the raids accomplished was unclear. They reflected Bush's unwillingness to deal decisively with Iraq then.

Baghdad continued to deny the UN weapons inspectors unimpeded access. Additionally, it activated targeting radar in the no-fly zone in northern Iraq. The administration debated how to respond. Chairman of the Joint Chiefs of Staff General Colin Powell, as well as several other high-ranking Pentagon officials, wanted to reply with a major assault. But Bush chose a minimal response. On January 17, the United States attacked the Zafaraniya manufacturing complex outside Baghdad with cruise missiles, a move intended to reinforce the UN weapons inspectors' demand for unconditional permission to fly into Iraq. An errant missile hit Baghdad's Rashid Hotel, where a conference of Islamic radicals was being held.

Baghdad relented on the weapons inspectors, but Iraqi challenges in the no-fly zones continued. Another U.S. attack on Iraqi air defense installations followed on January 18, intended to destroy targets that had been missed in the raid five days before.

On January 19, Saddam announced a "unilateral cease-fire," as Bush had done some two years before at the end of the Gulf War. Thus, it was not Bush who had called the last cease-fire in the Mother of Battles; it was Saddam. Saddam presented his "unilateral cease-fire" as an overture to Bill Clinton, who was to be sworn in as president the next day.

In the last days of the Bush administration, Clinton waited in Little Rock, Arkansas, for Inauguration Day. There, he misread the dynamics of the military clashes occurring between the United States and Iraq. Clinton believed that Bush, rather than

Saddam, was responsible for the conflict, picking a fight he meant to leave to Clinton.

In the midst of the escalating American-Iraqi tensions, President Clinton gave an interview on January 13, 1993, to the *New York Times* in which he said that he was ready for a fresh start with Saddam. He explained that if Saddam were sitting on the couch next to him, Clinton would advise him, "If you want a different relationship with me, you can begin by observing the UN requirements and change your behavior. I am not obsessed. . . . I am a Baptist. I believe in deathbed conversions." Clinton further suggested, "If [Saddam] spent half, maybe a third of the time worrying about the welfare of his people, that he spends worrying about where he positions his SAM missiles and whether he can push the boundaries of the cease-fire agreement, I think he would be a stronger leader and in a lot better shape over the long run."[6]

The next day Clinton reversed himself, affirming that there would be no difference between his policy and Bush's. That was just as well. Some 74 percent of the American public wanted Clinton to continue Bush's policy toward Iraq, while 55 percent opposed normal relations with Iraq, even if Saddam changed his behavior and abided by the UN resolutions, as Clinton had proposed.[7]

Still, doubts persisted about what Clinton's real policy would be. After all, the top decisionmaker himself had suggested that he was prepared to come to terms with Saddam. And even after Clinton's disavowal, the new administration kept insisting that it was "depersonalizing" the conflict with Iraq. What else could that mean, but that there were circumstances under which it could come to terms with Saddam?

The Clinton administration's initial formulation of its Iraq policy produced such headlines as "U.S. Drops Demand for Saddam Ouster" and "Clinton's Effort to Depersonalize Policy toward Iraq Makes Arab Allies Nervous."[8] The administration's policy did not begin to clarify itself until Martin Indyk, National Security Council adviser on the Middle East, articulated the policy of "dual containment" in May 1993.[9] It became apparent that the new administration was no more inclined to lift sanctions than was its predecessor. Still, any dispassionate, objective analysis of the Clinton administration's position in its first few months would have suggested that Iraq stood a reasonable chance of getting sanctions lifted.

In fact, Baghdad initially took a positive view of the new administration, or so it seemed. Even during the presidential campaign, the Iraqi press explained that Clinton, if elected, would be more interested in dealing with domestic problems and less inclined to continue the confrontation with Iraq.[10]

Baghdad announced its January 19, 1993, "unilateral cease-fire" as a "goodwill gesture" to Clinton. Six days later, Deputy Prime Minister Tariq Aziz explained that Iraq wished to open "a new chapter with the new administration," while urging the Clinton team not to "personalize the conflict." On February 14, Saddam met with Ramsey Clark, attorney general under President Lyndon Johnson, who in subsequent years had gone on to distinguish himself by praising Ayatollah Khomeini as a "saint" and by writing a book about American war crimes in the Gulf War.

In his meeting with Ramsey Clark, Saddam praised Clinton's opposition to the Vietnam War and advised that if Clinton wanted to be reelected in 1996, he should concentrate on the economy and "avoid any large military burdens." Saddam offered an olive branch, explaining, "I believe that the President of the biggest country in the world . . . needs to try wisdom and not weapons," and suggesting that we "lay the groundwork for new relations based on mutual legitimate interests and mutual respect, leaving the past behind us."[11]

Less than two weeks later, the bomb under the North Tower of the World Trade Center exploded. Saddam's meeting with Clark seems to have been a reprise of old tricks, designed to mislead his opponents. A week before invading Kuwait, Saddam had asked to see the U.S. ambassador to Baghdad, April Glaspie. He managed to convince her that even with 100,000 Iraqi troops on the Kuwaiti border, his intentions were benign. Glaspie sent back to Washington a cable entitled "Saddam's Message of Peace," and the White House chose to underscore Glaspie's conciliatory posture with a message of its own, sent under President Bush's name: "My Administration continues to desire better relations with Iraq."[12]

Saddam, almost certainly, had resolved on invading Kuwait long before he received the Bush administration's conciliatory messages in July 1990. Similarly, Saddam had decided on revenge long before Clinton's conciliatory statements of January 1993. But in each case, the olive branch offered by the president or president-elect only encouraged Saddam to continue with his plans, rather than to reevaluate them and back down.

If Saddam were a more conventional figure, he would have tried to exploit the fact that there was a new U.S. administration and that the new president, shortly before he took office, had indicated that he was prepared, under the proper circumstances, to come to terms with Iraq. Such a political leader would have cooperated for the period necessary to assure the new U.S. administration that a deal could be reached with Baghdad. Indeed, Washington's Iraq experts explicitly warned the Clinton administration at that time: "Beware of Saddam's charm campaign." The expectation was that Saddam would comply minimally with UNSCOM, get sanctions lifted, and then go back into the business of manufacturing proscribed weapons.[13]

But there was no charm campaign. Instead, Saddam proceeded with a campaign of revenge. Indeed, Saddam has long been using violence and force, and the threat thereof, to divide the coalition against him. He has acted far more systematically than has generally been recognized. Almost every year since 1993, he has sought to achieve some objective that is related to undermining an aspect of the post–Gulf War constraints on Iraq. Should it be any surprise, then, that in the course of that effort, Iraq occasionally carried out acts of terrorism? We shall discuss those acts of terrorism, or apparent acts of Iraqi terrorism, in chapters 17 and 18.

In this chapter we will look at how Saddam has been coming back. He seems to think in one- or two-year time periods. He is fixed and focused on a specific objective. And he has been coming back, *not* by complying with the provisions of the Gulf War cease-fire, but by defying them.

Saddam's Post–Gulf War Comeback—The Years before Hussein Kamil's Defection

Following the Gulf War cease-fire, Saddam spent the next two years consolidating his internal position. And by early 1993, he was ready for a campaign of terrorism. More incidents followed the Trade Center bombing. They included the attempt to assassinate George Bush and his entourage in April. And when, in late June, Clinton launched a cruise missile strike on Iraqi intelligence headquarters, the attack temporarily ended Saddam's campaign of revenge. But as a prominent Iraqi journalist wrote then, "It is a wrong conclusion on Clinton's part that striking the intelligence headquarters will ease the Iraqis' anger. This only confirms Clinton's weakness and naïveté."[14] In other words, Saddam would be back.

1994—The Lunge at Kuwait. If the focus of Saddam's efforts in 1993 was terrorism, it was something different in 1994. Iraqi threats, beginning in January, culminated in an Iraqi lunge at Kuwait in October. That move was more complicated and nuanced than many people recognized at the time. And it was foreshadowed by Iraqi statements made much earlier that year.

In a January 16, 1994, speech marking the third anniversary of the start of the Gulf War, Saddam threatened the Gulf states directly and America indirectly:

> For all evildoers, masters and slaves, we reiterate that they must not be deluded once again or miscalculate things. . . . The punishment of the criminals is an eye for an eye and a tooth for a tooth. . . . For all those concerned among the Arab rulers, we say: All that you have done against the people of Iraq is something that angers God. . . . I also say that there is still some way out of this. I tell the wise believers: Will you do so? I have conveyed the message, and may God be my witness."[15]

Patience is a key virtue for "the believer" in Islam. A few days later, the number two man in Iraq's Information Ministry, Nuri Najm al-Marsoumi, published an article entitled "Beware of the Patient Man's Rage." Al-Marsoumi wrote, "We will not allow the agents, lowly people, or covetors to encroach on what we own, on our rights, and on our dignity. As the leader said, the punishment of the criminal will be an eye for an eye and a tooth for a tooth. The Arabs say, 'Beware of the patient man's rage.' "[16]

In March 1994, Tariq Aziz visited New York to lobby the Security Council before a mid-month sanctions review. Shortly in advance of his trip, Saddam addressed the Iraqi cabinet:

The blockade has been continuing for four years. . . . Every day one commission comes and another one goes. They have destroyed the weapons, the plants, and the factories they wanted to destroy. . . . How long will the Iraqi people wait, and how can they accept the diplomacy of the fox of the jungle [Ekeus]? . . . It is *their right to choose alternatives* if they discover that this way will not lead to a positive result. . . .

If comrade Tariq Aziz returns without obtaining the Security Council's agreement to meet its obligations and if there is no hope that the injustice will be lifted from the Iraqi people, then the Iraqi people and their leadership can do nothing else *but decide what they believe will give them hope, God willing, in the direction they believe is sound.*[17]

A press campaign followed, elaborating on Saddam's statement. As the Iraqi News Agency reported,

Al Thawrah and *al Qadisiyah* write that President Saddam Hussein's speech to the Cabinet on March 13 has ended a period of wily procrastination and provocation and put the onus on the UN Security Council to discharge its clear and pressing obligations. . . . The Council should uphold justice and fairness and fulfill its obligations by ending injustice and lifting the sanctions, *or else Iraq and its leadership would legitimately choose a new road, a road different from the one it has followed for the past four years.*[18]

What did it mean to "choose a new road?" What "alternatives" did Iraq have? On July 17, in a speech marking the twenty-sixth anniversary of the coup that brought the Ba'th Party to power, Saddam again threatened the Gulf states: "We reiterate that we offer peace and security to whomever needs them, including rulers who harmed us. . . . We have offered what is satisfactory to God and freed ourselves from blame. . . . God be my witness that I have delivered the message."[19]

Blame for what? In September, a large Iraqi delegation was in New York for the opening of the UN General Assembly. It included General Amir Rashid, then director of Iraq's Military Industrial Organization. On September 26 he left for Baghdad, leaving the rest of the delegation behind. UNSCOM's next report was due to come out on October 10. But UNSCOM was not quite ready to give Iraq a clean bill of health. And Rashid was unhappy with what UNSCOM had told him.[20]

The next day, September 27, Saddam gave a speech, broadcast on Iraqi TV. Saddam looked listless throughout most of the speech, but he grew energized toward the end, when he proclaimed,

> When your *patience* or the *patience* of the Iraqis comes to an end, because of this embargo, with our own arms, we can open the storehouses of the universe and feed you until you are full of rice, flour and sugar and you know what I mean.
>
> When the *patience* of the Iraqis comes to an end because of this embargo and they begin to murmur and they are hungry, then, by God, we will open to all the storehouses of the universe and anyone who hears this should say that Saddam Hussein said this.

Two days later *Babil* elaborated on Saddam's speech:

> The great leader Saddam Hussein has said, "When we feel that the Iraqis might starve, by God, we will open for them the storehouses of the universe." The United States must understand that there is *a limit to the patience of the Iraqis*. They must not imagine that the destruction of Iraqi missiles and other weapons will strip the Iraqi people of their ability to influence matters and change the course of history. The will of people who want to live in dignity is the strongest and most effective missile.
>
> Does the United States realize the meaning of opening the storehouses of the universe with the will of the Iraqi people? . . . Does it realize the meaning of every Iraqi becoming a missile that can cross to countries and cities?[21]

The Iraqi press soon began to threaten terrorism on a massive scale: "The phase of Arab and Muslim *patience* is about to reach an end and . . . souls have begun to prepare as of now for hot confrontations in more than one country and continent." "History says that when peoples reach the verge of collective death, they will be able to spread death to all." "We seek to tell the United States and its agents that the Iraqi *patience* has run out and that the perpetuation of the crime of annihilating the Iraqis will trigger crises, whose nature and consequences are known only to God."[22]

Then suddenly, on October 7, in a noontime press conference, Clinton announced that 60,000 Iraqi Republican Guards were headed toward the Kuwaiti border and that he was rushing U.S. forces to the region. Had Clinton not done so, Iraqi troops probably would have crossed the border, taking Kuwait or a part of it. Then Saddam would probably have turned around and offered to return what he had seized in exchange for the lifting of sanctions.

But that did not happen. And most probably Saddam did not expect it to happen. Most probably, Saddam expected that the United States would respond as it did. A Western diplomat later characterized Saddam's strategy by saying, "You create a crisis, aggravate the crisis, and you have changed the dynamics in your negotiations."[23] Indeed, *Washington Post* columnist Jim Hoagland cautioned at the time, "Do not be surprised, or deceived, if Saddam now indicates that he will lower the tensions he has created and recognize Kuwait's frontier, the major remaining hurdle he has to clear to get sanctions lifted."[24] And that is exactly what happened. The next month, under Russian mediation, Iraq formally recognized Kuwait, removing what seemed to be pretty much the last obstacle to lifting sanctions.

1995—Eliminating UNSCOM? In late 1994, Ekeus told his staff that he was moving to give Iraq a clean bill of health. Some members of UNSCOM believed that Iraq had a biological weapons program it had not declared. Ekeus advised that they would have to come up with hard evidence soon.

As the new year began, Saddam, addressing a group of army officers, asked, "Who will fire the fortieth missile?"—a reference to the missiles that Iraq fired on Israel during the Gulf War.[25] Baghdad also announced that 1995 would be the year of lifting the sanctions. And at the time, it seemed that Iraq might well prove correct. But two things happened to prevent that turn of events.

First, the U.S. position hardened. The Clinton administration had given thought to a partial lifting of sanctions. But the notion was blasted by Republicans, who had just taken control of both houses of Congress in the 1994 elections. The idea was dropped. In March 1995, for the first time, the administration said that it would veto any proposal to lift sanctions.

Second, the UNSCOM staff, who believed that Iraq had an undeclared biological weapons program, obtained the evidence to prove it. In early 1995, UNSCOM received detailed information about Baghdad's import of very large quantities of biological growth material from Europe. The quantities were said to be "mind-boggling."[26] As UNSCOM began asking the Iraqis what they had done with that material, the Iraqis could provide no coherent answer.[27] The Iraqi press began to growl. In mid-March, *Babil* advised, "The matter effectively requires the setting of a deadline for the continuation of diplomatic negotiations."[28]

UNSCOM came under increasing pressure from Iraq's friends on the Security Council—France and Russia—to declare that it had taken care of Iraq's weapons programs. Tariq Aziz visited New York in early March to press Iraq's case. UNSCOM's next report was scheduled for April 10. That report was so important to Iraq that Aziz returned to New York to be there when it was issued.[29] And because of the questions UNSCOM had asked, the Iraqis were well aware that UNSCOM might report that Iraq had a proscribed, undeclared biological weapons program.

And that is what happened. On April 10, UNSCOM formally stated for the first time that Iraq had such a program. The next day, Nuri al-Marsoumi, who had written "The Patient Man's Rage," published an article entitled "A U.S. Misunderstanding which Must be Removed." Al-Marsoumi cautioned, "Iraq's abandonment of part of its weapons—the long-range missiles and chemical weapons— . . . does not mean that it has lost everything. . . . The Iraqi people consist of 18 million people. . . . Should it be necessary, the people can become a huge potent force in defense of their own interests."[30]

In subsequent years, it became clear just how much Saddam valued Iraq's biological program. From UNSCOM's April 10, 1995, report until December 16, 1998, when UNSCOM's presence in Iraq ended, Baghdad never turned over any of its biological stockpile. It always claimed that it had unilaterally destroyed that material. And until the end, Iraq's biological program remained the most elusive and mysterious of its proscribed unconventional weapons programs—the one that Baghdad took the most pains to conceal from UNSCOM. And because it seems to have always been Saddam's intent to retain Iraq's entire biological weapons program, UNSCOM's April 10, 1995, report raised the prospect that sanctions would never be lifted, or at least not as envisaged in UNSCR 687.

Yet that was not so clear at the time. It seemed only that the April 10 report had slowed the momentum that had been building to lift sanctions. Instead of being close to declaring that it had taken care of all of Iraq's proscribed weapons programs, UNSCOM suddenly had an entirely new program to deal with.

Baghdad insisted that it would not discuss its biological program until UNSCOM cleared Iraq on its chemical and missile programs. Under pressure from France and Russia, UNSCOM's next report, issued on June 20, did essentially that, or came very close to doing so. It said that "in the ballistic missile and chemical weapons areas, the Commission is now confident that it has a good overall picture of Iraq's past programmes and that the essential elements of its proscribed capabilities have been disposed of."[31] That left only the biological program.

After four years of denying it had ever developed an offensive biological weapons program, in early July 1995, Iraq suddenly acknowledged that it had. On July 1, Ekeus led an UNSCOM team to Baghdad. Iraqi officials told UNSCOM that Iraq had produced more than 5,000 gallons of botulinum, and it had made nearly 160 gallons of anthrax. But they claimed that Iraq had never weaponized those agents. Rather, they maintained that the material was destroyed in October 1990, as war loomed, to prevent an accidental release in case the United States attacked storage and production sites. Yet aside from Iraq's belated admission that a biological program had existed, UNSCOM believed almost nothing of what the Iraqis said. Among other things, why would Iraq have produced such a large quantity of biological agents if it did not intend to put the material into weapons?

On July 16, Saddam met with U.S. Representative Bill Richardson, then a seven-term Democratic congressman from New Mexico. Richardson was also a close associate of Bill Clinton's and he had gone to Baghdad to bring back two Americans who had strayed across the Iraqi-Kuwaiti border and been imprisoned in Iraq. Of course, it was Iraq that arranged the trip and determined its timing.

Richardson's appearance in Baghdad represented a rare opportunity for Saddam to speak with someone close to top U.S. decisionmakers. And during his meeting with Richardson, Saddam would have tried to understand as clearly as he could the U.S. position on Iraq. Most likely, Richardson would have told him of the U.S. determination to maintain sanctions, at least until Iraq complied with UNSCR 687.

That night, Iraq announced a change of defense ministers. Saddam's cousin, Ali Hassan al-Majid, notorious for his genocidal suppression of the Kurds, was replaced by General Sultan Ahmed, a professional soldier, who had led the Iraqi delegation at the cease-fire talks that ended the Gulf War. Only once before had Saddam appointed a competent soldier not related to him as defense minister—from December 1990 to April 1991.

The next day, July 17, in his annual national day speech, Saddam declared that Iraq would cease its cooperation with UNSCOM if no progress were made toward lifting sanctions. But he gave no deadline. In Cairo a few days later, Iraq's foreign minister, Muhammad al-Sahaf, did give a deadline—August 31. On August 4, Ekeus again traveled to Baghdad, in an attempt to clarify some of the outstanding issues regarding Iraq's biological program. But Baghdad provided no new information. Instead, Tariq Aziz told Ekeus that Iraq would cease its cooperation with UNSCOM on August 31 if no progress were made. Immediately upon Ekeus's return to New York, Nizar Hamdoon, Iraq's UN ambassador, delivered a message from Aziz. On August 7, Hamdoon told Ekeus that Aziz had thought Ekeus might not have believed what he had said about the deadline. But Aziz wanted Ekeus to understand that he was serious and to report Iraq's demand to the Security Council.[32]

And then, on August 8, Saddam's son-in-law Hussein Kamil, who had supervised Iraq's unconventional weapons programs, defected to Jordan, claiming that Saddam intended to invade Kuwait and even the oil-rich eastern province of Saudi Arabia.[33] The United States again rushed forces to the region, amid suspicious Iraqi troop movements. Much later, Iraqi vice president Taha Yasin Ramadan offered another explanation. He said that the Iraqi leadership had decided to expel UNSCOM, but "Kamil's defection changed plans and compelled the Iraqi leadership to administer the battle in another direction."[34]

11

The Defection of Hussein Kamil—Iraq's Unconventional Capabilities

With Hussein Kamil's defection, Rolf Ekeus planned to travel to Jordan to meet with Kamil and learn what he might about Iraq's weapons programs. But Baghdad made clear that it would regard that meeting as an unfriendly act. It canceled the deadline it had just given Ekeus and invited him to Iraq instead. There, on August 17, 1995, Tariq Aziz explained that Hussein Kamil—on his own initiative, and unknown to the Iraqi government—had run an operation to conceal key Iraqi weapons programs from UNSCOM. That, of course, was nonsense. There was such an operation—but Saddam was behind it. And the decision to undertake it had been made back in April 1991, shortly after Iraq accepted UNSCR 687.[1]

In the following days, Ekeus heard Iraqi officials acknowledge not only that Baghdad had weaponized biological agents, as UNSCOM had suspected, but that it had done a number of other remarkable and dangerous things. Most dramatic was Baghdad's response to the Security Council's vote on November 29, 1990, authorizing military action against Iraq. Iraq took twenty-five SCUD missile warheads and filled them with the biological agents anthrax and botulinum. The missiles were taken out to distant airfields, where the commanders had instructions to fire them at targets in Israel and Saudi Arabia if, in the war that was to come, the coalition were to march on Baghdad.

In 1994 Frederick Forsyth published a best-selling thriller, *The Fist of God*, a fictional account of the Gulf War. The protagonist, Terry Martin, a brilliant young Arabist, serves as a consultant to the British government. In the novel, a U.S. spy satellite picks up the phrase "the fist of God" spoken in a Baghdad telephone conversation during the build-up to the war. It is not clear who spoke the words, but Martin becomes fixated on them, convinced that they refer to a doomsday weapon. British authorities pooh-pooh the idea, dismissing Martin as a "fusspot." Yet one official does not. Persistently sleuthing, the two come to understand that Saddam has a crash program to build a nuclear bomb, which he is ready to use, if coalition forces go beyond the liberation of Kuwait and actually enter Iraqi territory. Just before the allied offensive into the Iraqi desert is to begin, they manage to persuade senior British officials that they are right. The offensive is postponed. Martin's derring-do brother, a Special Air Services officer operating covertly in Baghdad, finds and demolishes the bomb. And the coalition armies go on to finish the war.

What is fact and what is fiction? During the Gulf War, as it turns out, Saddam did indeed have a doomsday device. But it was biological, not nuclear. And the trigger for its use was the fall of the regime, not the coalition's entry into Iraq. After all, what does Saddam care about Iraq's southern desert?

Strange as it may seem, Saddam believes he is bringing glory to the Arabs after a millennium of decline. And since, in his view, life is without worth if it does not include glory, he is justified in risking the lives of Iraqis in his repeated military adventures, while taking the lives of his enemies. Terry Martin explains Saddam's mindset:

> If [Saddam] can hurt America, he will win. If he can hurt America badly, really badly, he will be covered in glory. . . . The terror that revolts us has no moral downside for him. The threats and bluster make sense to him. Only when he tries to enter our world—with those ghastly PR exercises in Baghdad, ruffling that little boy's hair, playing the benign uncle, that sort of thing—

only when he tries that does he look a complete fool. In his own world he is not a fool. He survives, he stays in power, he keeps Iraq united, his enemies fall and perish.

"Terry, as we sit here, his country is being pulverized," a British official protested. "It doesn't matter, Simon," Martin replied. "It's all replaceable."[2]

In reality, immediately after the Gulf War, Uriel Dann, a professor of history at Tel Aviv University, wrote similarly,

> In post–Persian Gulf War analysis, one critical point has not been stressed sufficiently. This is Saddam's personality and its significance for the future so long as he is in Baghdad. Again and again, the man has been called a thug, a mafioso, a ruthless dictator, crazed with blood-lust, drunk with self-love, devoid of all humane impulse. All of this is deserved. Less tribute has been paid to his intelligence, his cunning, his adaptability, his perseverance, his coolness under duress—and to his leadership. Indeed, few have noted the heroic measures of these qualities. . . .
> Saddam Hussein does not forgive and forget. His foes brought him close to perdition and then let him off. . . . He will strive to exact revenge as long as there is life in his body. . . . And when he does hit, he may, by the grace of God, miscalculate as he has miscalculated in the past. But even so the innocent will pay by the millions. This must never be put out of mind: Saddam Hussein from now on lives for revenge.[3]

Saddam's Retained Unconventional Capabilities

Hussein Kamil's defection precipitated a flood of stunning revelations about the proscribed unconventional capabilities that Iraq had retained after the Gulf War and managed to conceal from UNSCOM and the International Atomic Energy Agency (IAEA), responsible for dismantling Iraq's nuclear program. Iraq had systematically turned over the least significant elements of its weapons programs, while keeping those that were most dangerous.

Iraq's Nuclear Program. Like life imitating art, it emerged after Kamil's defection that Saddam had indeed initiated a crash program to build a nuclear bomb, following his invasion of Kuwait. Moreover, Saddam wanted a bomb that could be delivered by a SCUD missile, not just a nuclear device that could be exploded in the desert to scare people. Nor did Saddam want a bomb that could be dropped by an airplane. That would have been easier for Iraqi scientists to build, because its size would not have been so important. But an airplane was not a practical means of delivery. Saddam wanted a bomb he could use.

And by November 1990, Iraq's nuclear scientists had managed to assemble a nuclear device, minus the uranium core. They planned to take the fissile material for the core from the fuel in Iraq's French and Russian research reactors. But they still faced a major problem. Their design for the explosives that compress the uranium core to a critical size—necessary for a nuclear reaction to occur—was too large. Thus, they did not quite have it. They could not then make a bomb small enough to be carried by a SCUD missile. And the Gulf War soon began. It shut down Baghdad's electrical grid and otherwise interrupted Iraq's nuclear program, ensuring that Saddam did not develop a bomb at that time.

But Iraq's bomb program continued *after* the war, as was learned after Kamil's defection. While most Iraqi scientists were pulled from their regular work to assist in postwar reconstruction, the bomb design team was kept together and kept on working. In the four years between the war's end and Kamil's defection, the scientists solved the problem that had stumped them before the war. They came up with a design for a nuclear bomb small enough to fit on a SCUD missile. It was 32 to 35 inches in diameter and weighed one ton.[4] Thus, Iraq's position in August 1995 was the reverse of what it had been in January 1991. Then it had possessed the fissile material for a bomb, but lacked a suitable design. In August 1995, Iraq had a suitable design, but lacked the fissile material.

Meanwhile, the collapse of the Soviet Union and the political and economic disarray in Russia created a potential opportunity for Baghdad to acquire fissile material on the black market there. That was a concern, even while weapons inspectors were in Iraq, as it seems that Iraq needs to obtain only 35 pounds of highly enriched uranium to build a nuclear bomb.[5] But as long as the weapons inspectors were in Iraq, they provided some check on

Iraqi activity. If Iraqi scientists tried to build a bomb while inspectors were there, the possibility existed that a surprise inspection could catch them. Moreover, UNSCOM's monitoring system kept track of equipment being used for civilian purposes that could also be used in Iraq's weapons programs. Among other things, shaping a nuclear bomb core requires high-precision machinery. Such equipment was regularly tagged and checked to ensure that it was not diverted for proscribed purposes, while UNSCOM was there. Of course, once UNSCOM left Iraq, that was no longer possible.

Subsequently, the prospect that Iraq might actually produce one or more nuclear bombs was, or should have been, a much heightened concern. Indeed, in the spring of 1999 Ekeus, no longer UNSCOM chairman but now Sweden's ambassador to Washington, warned, "Saddam will get a bomb, because these materials are floating in. Every day, they are more advanced."[6] Khidhir Hamza, a senior Iraqi nuclear scientist who defected in 1994, agreed. Hamza estimated that once Iraq acquired the fissile material, it would have a bomb two to six months later, depending on the form in which Iraq obtained the material and how long it took to reshape it into a nuclear bomb core.[7]

Iraq's Chemical and Missile Programs. Kamil's defection also brought to light important information about Iraq's chemical weapons program, particularly its capability to produce VX. VX is a highly lethal chemical agent. It is fatal not only when inhaled, but also when absorbed through the skin. Unlike most chemical agents, VX does not dissipate quickly. It is sticky and viscous. And 1/100th of a gram is fatal. A person has only to touch a surface that has been struck with VX and he will die. VX is particularly useful as a way to deny equipment and territory to an enemy. For example, if Iraq were to attack U.S. military equipment pre-positioned in Kuwait with VX, U.S. forces would not be able to use that equipment until after it had been very carefully scrubbed down—that is, after Iraqi forces had overrun their target.

Iraq had long denied to UNSCOM that it produced VX. But following Kamil's defection, it was learned that Iraq had indeed done so. Iraq then acknowledged having produced 3.9 tons of VX, but claimed to have destroyed the material unilaterally.

It also turned out that Iraq had succeeded in indigenously producing SCUD missiles. That threw into doubt UNSCOM's belief that Iraq's SCUD missiles had been destroyed, a belief based on the assumption that the only SCUDs Iraq had ever possessed had been imported from the Soviet Union. Russia provided UNSCOM with a detailed accounting of Soviet SCUD exports to Iraq. So it seemed that all UNSCOM had to do was to count the number of missiles destroyed in Iraq after the Gulf War, along with those Iraq had used during the Iran-Iraq war, and when that figure matched the number of SCUD missiles Moscow had sold to Baghdad, then UNSCOM would have destroyed all of Iraq's missiles.

But the postwar destruction of Iraq's missiles was more complicated than it was supposed to have been. According to UNSCR 687, the weapons inspectors were to supervise the destruction of all of Iraq's proscribed weapons programs. UNSCOM destroyed some Iraqi SCUDs. But Iraq unilaterally destroyed others, in violation of UNSCR 687. To substantiate its claim that it had really destroyed the missiles it said it had, Iraq invited UNSCOM to count the missile engines at the sites where the destruction was said to have occurred. But once it was learned that Iraq had itself produced missile engines, the exercise became useless. Who could tell whether the wreckage from the missiles Iraq claimed to have destroyed came from Soviet SCUDs or indigenously produced ones? General Wafiq Samarrai headed Iraqi military intelligence during the Gulf War. He defected in late 1994, claiming that Iraq still had at least eighty SCUD and al-Hussein missiles—modified SCUDs, with a longer range.[8] Samarrai was not believed then. But he gained new credibility after Kamil defected.

Iraq's Biological Program. Because Iraq's biological program was so large and the efforts it had made to hide the program so extensive, this was perhaps the most awkward program for Iraq to explain to UNSCOM. But explain Iraq did—sort of.

Iraq's information minister did not normally attend meetings with UNSCOM. But he was the senior official seated opposite Ekeus as Iraq presented the new information on its biological warfare program. The minister, Abdul Ghani Abdul Ghafur, had the authority to compel a command performance from anyone UNSCOM wanted to see. And the meeting took on an eerie, ritualistic routine. As Ekeus explained, he would ask Abdul Ghafur a question.[9] Abdul Ghafur would reply that he did not know the answer, but he would bring in someone who did. A "peculiar parade" of scientists subsequently appeared.

It was late at night. The police waited outside. They would be given a person's name and they would go to his house and tell him to come with them immediately. One man later told UNSCOM he thought he was going to be executed, so he bade farewell to his wife and children.

And after the scientist was brought in, Abdul Ghafur would turn to him and ask a question on behalf of UNSCOM, such as, "Tell us what you've been doing on the virus research." Of course, that and many other issues discussed that night were important and sensitive subjects. The scientists did not know what to say. They had had no time to prepare, and they knew they were not supposed to reveal state secrets. They were scared, even shaking. And they would lie, and trick, and blink at the lights.

And so it was learned that Iraq had produced much larger amounts of anthrax and botulinum than it had acknowledged only the month before, and had produced them over a much longer period of time than it had admitted. Also, Iraq had produced a number of other biological agents, including ricin, a highly lethal toxin extracted from castor beans; clostridium perfingens, which causes gangrene; and aflatoxin, which causes liver cancer. Iraq also acknowledged having experimented with exotic and deadly viruses related to ebola, as well as camel pox, which many in UNSCOM, including Ekeus, suspected was a cover for smallpox-related work.

It also emerged that Iraq had continued to produce biological agents *after* the Gulf War. In May 1996 UNSCOM destroyed Iraq's biological production facilities, as they had been revealed after Kamil's defection.

It also turned out that Iraq had developed spray tanks that could be attached to manned aircraft or drones (unmanned aircraft) to disseminate biological agents. UNSCOM considered them "the most efficient means for the delivery of biological warfare agents produced by Iraq."[10] They were targeted during Operation Desert Fox, the four-day U.S.-British bombing campaign of December 1998. British defense secretary George Robertson described the Iraqi project, underscoring that the development of Iraq's unconventional weapons programs never stopped—not with the Gulf War, and not even with Kamil's defection. As Robertson explained,

> In 1990, Saddam ordered the production of unmanned aircraft to spray chemical and biological agents on civilians and ground troops that he might wish to attack. Early efforts to convert combat aircraft were not successful, but spraying equipment was successfully tested using an anthrax-like substance. In 1995, Saddam launched a new programme using a converted training aircraft code-named L29. The first flights were started in 1997 and the testing programme is still continuing. This aircraft has been fitted with two under-wing weapon stores capable of carrying 300 litres of anthrax or other nerve agents. If this were to be sprayed over a built-up area such as Kuwait City, it could kill millions of people. Once perfected, we suspect that Saddam had intended to deploy these drones of death in Southern Iraq as a direct threat to his neighbours, and it was this development programme that we hit.[11]

Of course, that program is up and running again. No one would seriously argue that four days' bombing permanently ended it.

Also, in the period following Kamil's defection, UNSCOM came to believe that Iraq had tested biological agents on human beings. UNSCOM found two human-size inhalation chambers for testing the lethality of biological and chemical agents. Iraq said it had used animals, such as donkeys, in the chambers, "but inspectors note that they are primate-shaped and that Iraq did not use monkeys to test germ or nerve weapons."[12] Former inspector Scott Ritter explained UNSCOM's view in some detail:

> Iraq undertook a program, run by the office of the president and involving a special MIC (Military Industrial Commission) Unit 2001, either to produce new agent, or test agent that was retained from pre-Gulf War stocks. In 1995, Unit 2001 conducted tests on live human subjects taken from Abu Ghaib prison, using BW and binary CW agent. Around fifty prisoners were chosen for these experiments, which took place at a remote testing ground in western Iraq. The purpose of these experi-

ments was to test the toxicity of available agent to ensure that the biological arsenal remained viable. As a result, all the prisoners died.[13]

Yet despite all that UNSCOM learned about Iraq's biological program after Hussein Kamil defected, there was much that UNSCOM never learned. This program continued to be the one that Baghdad made the most effort to conceal. Just before Ekeus left Iraq after his stunning visit there in August 1995, he was taken to a place that Iraqi authorities claimed was a chicken farm, belonging to Kamil. On the premises was a stash of some 600,000 documents relating to Iraq's unconventional weapons programs in hundreds of boxes. There were, for example, 100 boxes dealing with Iraq's nuclear program. But there was only one box on Iraq's biological program.

Terrorism is among the very serious dangers posed by Iraq's biological program. A tiny amount of anthrax—just one gram—theoretically contains 100 million fatal doses. Anthrax is "100,000 times deadlier than the deadliest chemical warfare agent," according to the Defense Department.[14]

The most lethal way of disseminating anthrax would be with an airplane, to which was attached a spraying device. The idea would be to create a cloud of the biological agent that would be carried over a city. Official estimates of the fatalities that could ensue range from 100,000 to a few million.[15] A major part of the problem is that if an anthrax attack were to occur, authorities would not know it had taken place until people began to show symptoms. But at that point, there is no treatment. Some 85–90 percent of those who had been exposed to the anthrax attack would die, even if they were treated with massive amounts of antibiotics.[16]

But carrying out such an attack is difficult. The anthrax must be aerosolized, disseminated in a spray of particles. And it must be produced at a very small and specific size: between one and five microns (a millionth of a meter). If they are too big, the anthrax particles will be filtered out by the body's natural defenses. If they are too small, they will not settle in the lungs, but will be exhaled. The Japanese doomsday cult Aum Shinrikyo tried to spread anthrax and another biological agent some ten times between 1990 and 1995, but it was completely unsuccessful.[17]

Given the extreme secrecy that surrounds Iraq's biological program, one point is especially troubling. Of all Baghdad's unconventional weapons programs, its biological program would be the easiest to reconstitute. Iraq could have turned over to UNSCOM the vast bulk of the biological agents it produced, retaining only a small seed stock that could be quickly regrown, if Baghdad chose to do so. Why didn't Saddam do so?

Seth Carus, a highly regarded expert in the field, has offered one possible explanation. Biological agents, like all living organisms, contain DNA. If there were a biological terrorist attack and authorities had possession of part of the stockpile from which the agents used in the attack had come, these might raise suspicions as to who was behind the attack, through DNA testing. If they did not have any of that stockpile, however, that would be very difficult. And by retaining his entire biological stockpile, Saddam maintains the option for carrying out biological terrorism.

Saddam's Post-Gulf War Comeback, since Hussein Kamil's Defection

The defection of Hussein Kamil shook the Iraqi regime. Kamil had been a very important figure in Baghdad. Moreover, his defection prompted Jordan to turn against Iraq. King Hussein had backed Saddam during the Gulf War and had remained friendly afterward. But with Kamil's defection, the king was ready to work for Saddam's overthrow. He welcomed Kamil and the other Iraqi defectors who followed.

1995–1996—Recovering Internally and Eliminating the Iraqi Opposition. Saddam had been caught *in flagrante delicto*. The core of the Gulf War cease-fire, UNSCR 687, was the destruction of Iraq's unconventional weapons programs. Suddenly, Saddam had been shown to be in brazen violation of the cease-fire. Moreover, before Kamil's defection, Iraq had been generally considered little threat, essentially defeated and defanged. But with the defection, Iraq was revealed to be a very considerable threat.

A serious U.S. response should have followed, but there was none. The declared policy of the Clinton administration toward

Iraq had been "containment." And maintaining sanctions was a key element of that policy. Up until Kamil's defection, it had been a struggle to maintain a Security Council consensus in favor of sanctions. But once Kamil defected, that ceased to be a problem.

However, Kamil's defection also made it clear that Iraq was such a serious threat that a policy whose central focus was to maintain sanctions was inadequate. Sanctions did not address the danger posed by the unconventional capabilities Saddam retained. Kuwait and Saudi Arabia became especially concerned. Even Israeli officials raised questions about the U.S. handling of Iraq.

The administration's standard response to the expression of those concerns was that the United States was taking care of the problem.[18] By that it meant that it would overthrow Saddam. Thus, after Kamil's defection, the administration in effect had two policies. Its public policy remained "containment." But its real, undeclared policy, or so it told America's allies, was to overthrow Saddam in order to address the unconventional threat he was recognized to pose in the wake of Kamil's defection.

From the Bush administration, in fact, the Clinton administration had inherited two CIA programs to overthrow Saddam. One aimed at bringing about a coup. It involved working with a group of ex-Ba'thists and former Iraqi officials who claimed they could carry one out. The other CIA program involved trying to overthrow Saddam through a popular insurgency. That was to be led by an umbrella organization, the Iraqi National Congress (INC), which encompassed several broad-based opposition groups. Yet soon after assuming office, the Clinton administration halved the funds for covert activities in Iraq. It apparently believed that Saddam's overthrow, while desirable, was not necessary.

The coup option, while it might avoid prolonged civil strife and the need to develop a new form of government in Baghdad, had a fatal flaw. It was extraordinarily unlikely to succeed. Multiple and competing security organizations made Saddam's regime practically coup-proof. In fact, as the Bush administration came to recognize the difficulty of carrying out a coup, it turned to developing the alternative option of support for an insurgency.

Thus the INC had been established during Bush's last months in office, and it was promised U.S. support.

The Clinton administration appeared initially to support the INC and the strategy of supporting an insurgency that would overthrow the Ba'thist regime altogether, not just remove Saddam. In April 1993, a high-level INC delegation visited Washington. It met with National Security Adviser Anthony Lake and Secretary of State Warren Christopher. Vice President Al Gore himself even stuck his head into Lake's office to say hello during his meeting with the INC, as a gesture of the administration's support for the organization. The INC then proceeded to establish a base in northern Iraq, in territory controlled by the Kurds. There, the INC developed a network of covert ties with individuals in Iraq proper and looked forward to the time when it would lead an insurgency against Saddam. But that time never came.

Tellingly, the Clinton administration never supplied the INC with any arms. Nonetheless, in March 1995 the INC proceeded with a modest military operation. Its aim was to generate defections from Iraqi army units stationed in the north, near the Kurdish lines. But the White House panicked. A dubious opposition figure (who may well have been a double agent) flew to Washington to warn that the INC operation would draw the United States into a military confrontation with Saddam. Senior White House officials, including Lake, "became skittish and ordered the insurgency's offensive to a halt as it was about to be launched."[19]

Yet five months later, after Kamil defected, as U.S. policy shifted to overthrowing Saddam, the administration did not revisit its decision regarding the INC. Rather, it focused on the other option for overthrowing Saddam. It stepped up efforts to promote a coup.[20] Despite the experience of the Bush administration, senior officials in the Clinton administration believed that that would be easy. But they were proved quite wrong. In the summer of 1996, when Saddam had recovered from the domestic shock generated by Kamil's defection, he turned to eliminating the CIA operations against him. David Wurmser has described the fate of the CIA-backed coup attempt:

> The CIA's efforts climaxed in spectacular failure in July 1996, when an Iraqi intelligence officer located in Bagh-

dad contacted the CIA's station chief—over captured CIA communications equipment—and informed him that the Iraqis knew the detailed plans of the coup plot. The Iraqi informed the American that all the Iraqi officers and agents who had been involved in the plot had been rounded up and executed. . . . Gloating to the CIA station chief in Amman, the Iraqi told the Americans to pack their bags and go home.[21]

That left the INC as the only option for overthrowing Saddam. But instead of developing the INC, the administration dealt with it carelessly and soon lost it as well.

In the absence of any serious U.S.-backed effort to overthrow Saddam, the two Kurdish factions—Massoud Barzani's Kurdistan Democratic Party (KDP) and Jalal Talabani's Patriotic Union of Kurdistan (PUK)—had taken to fighting each other. Regularly, the INC would arrange a cease-fire. It would last for some time, but then the fighting would flare up again. Finally, following a meeting in November 1995 held under the auspices of the U.S. State Department, the Kurds agreed on terms for a cease-fire. They included provision for the establishment of an INC peace-monitoring force, to enforce the Kurdish cease-fire. The United States was to provide $2 million to equip it.

But the money never came.[22] The White House maintained that all the money for Iraq was in the CIA's budget. And that budget was meant for covert operations. But, the White House explained, the peace-monitoring force was an overt operation. Therefore, the CIA could not lawfully fund it. Of course, if the administration had really wanted to provide the $2 million, it could have found the money. The White House, for reasons it would not reveal, did not want to provide it. Perhaps it apprehended that if it provided arms to the INC for the peace-monitoring force, it would not be long before the INC again turned to fighting Saddam.

So the fighting between the Kurds continued and intensified. The PUK turned to Iran for assistance. And the KDP turned to Iraq. And in August 1996, at the invitation of the KDP, some 40,000 Iraqi troops marched north over an open plain, under the sunny blue skies of a Middle Eastern summer. The United States could see that those forces were planning to attack in the north and that the INC was their likely target. Despite U.S. promises to protect the INC, it did nothing. And over the Labor Day weekend, Iraqi Republican Guards assaulted the INC at its headquarters in Irbil.

The Clinton administration responded by proclaiming that U.S. interests lay in the south, rather than the north. Clinton extended the southern "no-fly zone" in Iraq one degree further north, to the suburbs of Baghdad. And he lobbed some cruise missiles at air defense sites in southern Iraq. Addressing the nation on September 3, Clinton explained what had happened: "It appears that one Kurdish group which in the past opposed Saddam now has decided to cooperate with him." And he maintained that "Repeatedly over the past weeks and months, we have worked to secure a lasting cease-fire between the Kurdish factions."

Of course, those who had followed the situation closely knew better. Iraq's assault on Irbil related only tangentially to the inter-Kurdish rivalry. It was really about eliminating Saddam's U.S.-backed opponents. Moreover, senior White House officials— against the advice of those in the bureaucracies—had been responsible for blocking the money for the Kurdish peace-monitoring force.

Nor did Clinton address the problem of how the United States would deal with the threat posed by Saddam's unconventional weapons, once it had tossed away its only remaining option for overthrowing Saddam. Indeed, Clinton's speech seemed to suggest that no such problem existed.

Baghdad's assault on the INC precipitated increased media interest in Iraq, resulting in at least one striking instance of administration "spin," or news management. *New York Times* columnist A. M. Rosenthal wrote a piece chiding the administration for allowing Saddam to retain dangerous unconventional weapons capabilities in violation of the Gulf War cease-fire.[23] Madeleine Albright, then UN ambassador, responded by calling up UNSCOM. Citing Rosenthal's column, she advised a senior official there, "We don't want to see any more articles like that. Why don't you stress the positive in your next report? Like all the material you've destroyed."[24]

If the administration believed that Iraq's unconventional weapons capabilities were not a significant danger, it should have said so. It might have claimed that, given continued sanctions, UNSCOM's presence in Iraq, and America's overwhelming military power, Saddam would never be able to use the unconventional capabilities he retained. And there could have been a public debate. Most probably, such a debate would have clarified the flaws in that position. After all, the administration itself had seemed to accept the opposite argument when U.S. allies expressed their concern about Iraq's unconventional capabilities as they became known after Kamil's defection, and it responded by saying that it would take care of the problem by overthrowing Saddam.

Thus, the administration held no clear and consistent view of the Iraqi threat and how it intended to address it. It all depended on circumstances. And by dealing with Iraq in that way, the administration was allowing a potentially dangerous situation to develop. Indeed, *Washington Post* columnist Jim Hoagland cautioned then of the policy disarray that followed from "the administration's growing inability to tell the world—and itself—the truth."[25]

Meanwhile, Saddam—having reestablished internal control, including the elimination of CIA operations intended to overthrow him—waited. Saddam waited to see which candidate would win the 1996 U.S. presidential election, and then he waited to see what the U.S. policy on Iraq would be in the next administration.

Of course, Clinton won the election. And Samuel (Sandy) Berger, who had been deputy National Security Council adviser during Clinton's first term, was named to replace Lake as NSC adviser. In December 1996, in his first extensive public comments on foreign policy, Berger described U.S. policy toward Iraq. He suggested that it was "a little bit like a Whack-a-Mole game at the circus: They bop up and you whack 'em down, and if they bop up again, you bop 'em back down again."[26]

12

The Expulsion of UNSCOM

In 1997, Saddam's focus was on UNSCOM. He sought to end UNSCOM's presence in Iraq. He moved incrementally. And by the end of 1998 he had succeeded, at minimal cost to Baghdad.

Saddam got rid of UNSCOM because he understood that it was not going to give Iraq a clean bill of health. Given the enormous amount of proscribed weapons capabilities Iraq retained, the chances were slim that Baghdad could persuade the Security Council to lift sanctions. Although Rolf Ekeus was set to leave his position as UNSCOM chairman at midyear, it was unlikely that the situation would change radically after his departure. Still, Saddam waited to see what Ekeus's successor would be like before making his first decisive moves.

In his last days as chairman, Ekeus repeatedly warned of the dangers posed by Saddam and his weapons. As he told one audience,

> The central question is what is the United States prepared to do regarding Iraq and its weapons of mass destruction. . . . As for the relation between Saddam and the international community, the situation is that the war is not over. If the situation is left unresolved, we will come to see the Gulf War as a brief parenthesis with no lasting significance. . . . We are nothing in Baghdad.

We are at their complete mercy. They can just stop our work at any time. The Security Council must increase its support for us. Basically, the United States must do so.[1]

But Ekeus's warnings fell on deaf ears.

Iraq's challenges to UNSCOM increased in Ekeus's last days. On June 10, Ekeus reported to the Security Council that on four separate occasions Iraqi officials had interfered with UNSCOM's operations. In the most egregious incident, the lives of all those on board an UNSCOM helicopter—UNSCOM staff, as well as the Iraqis accompanying them—were put at risk. As a weapons inspector on that helicopter related,

> The Iraqi minder unstrapped himself from his harness and nearly fell out of the aircraft. Clearly, they were under tremendous pressure to try to interfere with the mission. At one point the Iraqi [co]pilot who grabbed [the control stick] begged the [UNSCOM] pilot to leave the area, saying that "I have a wife and children" to worry about. Obvious implication: the man's family was threatened if he did not perform.[2]

At U.S. urging, the Security Council adopted a resolution (1060) on June 12, "deploring" Iraq's interference with UNSCOM and "demanding" that Iraq cooperate fully. It was to be the first of seven Security Council resolutions, culminating in UNSCOM's demise eighteen months later.

Indeed, the Clinton administration had never really provided the support for UNSCOM that the Bush administration had. Under Bush, when UNSCOM was prevented from carrying out an inspection, it would surround the site to which it had been denied access, waiting for the United States to gather support at the Security Council. And then, if Baghdad did not back down, the United States would threaten military action. To be sure, it was a cumbersome procedure, and on occasion even the Bush administration became distracted. Most egregiously, in July 1992 UNSCOM was denied access to the Ministry of Agriculture. It lay siege to that building for three weeks before it withdrew in frustration. A few days later it was allowed in—after Iraq had had time to clear everything out. Still, this less-than-satisfactory process enabled UNSCOM's work to go on.

Yet never once did Clinton do as Bush had done—demand that Iraq cooperate with a specific inspection, and use the threat of force to oblige it to do so. Rather early on, Anthony Lake advised Ekeus, "Don't give us sweaty palms."[3] That is, don't cause crises. And there were none during Clinton's first three years. But after Hussein Kamil's defection, UNSCOM became far more aggressive, and in return, Iraq blocked five inspections in March 1996. Yet each time, those inspections resumed within twenty-four hours.

That was not so two months later. In June, UNSCOM sought to inspect a Special Republican Guard site. The inspection team was blocked. UNSCOM responded by surrounding the site, maintaining a twenty-four-hour guard to prevent Iraq from sneaking anything out. In New York, Madeleine Albright began moving to have the Security Council declare Iraq in "material breach" of the cease-fire, using language that would prepare the way for the United States to threaten military action. But the White House felt that Albright was "moving too fast" and advised her to slow down. In New York, Ekeus alluded to "Tony Lake of the sweaty palms."[4] Without U.S. support, Ekeus ordered his team to stand down. He flew to Baghdad and reached an unsatisfactory understanding on procedures for inspecting so-called "sensitive sites."

In October, Tariq Aziz told Ekeus that there would be a split in the Security Council in 1997.[5] And, sure enough, the consensus that supported UNSCOM disappeared once Iraq began creating crises. In June 1997, as Iraq began what was to prove a systematic campaign against UNSCOM, the Clinton administration cast about for a way to add teeth to the Security Council resolutions. It came up with a proposal to ban the international travel of those Iraqi officials involved in blocking UNSCOM's work. But it did not propose banning the travel of high-ranking officials, like Aziz. That would have challenged a basic principle of national sovereignty, and the Security Council was not ready to support that. Thus, the proposed ban applied only to lower-ranking officials. But either those officials did not travel, or they could easily travel under another name. So the measure was essentially meaningless.

As Iraqi challenges to UNSCOM continued, on June 21 the Security Council approved another resolution (1115). It suspended the sixty-day sanctions reviews. Yet they had become irrelevant after Kamil's defection. Since UNSCOM was not going to give Iraq a clean bill of health, what did it matter whether sanctions were reviewed every 60 days or every 600 days? UNSCR 1115 also expressed the council's "firm intention" to impose "additional measures on those categories of Iraqi officials responsible for the noncompliance," unless UNSCOM said in its next report that Iraq was cooperating.

The following day, Iraq's Revolutionary Command Council (RCC), its highest governing body, and the Ba'th Party leadership met in joint session. They issued a defiant statement asserting that "the recent Security Council resolution contradicts the facts. . . . Iraq has complied with and applied all relevant resolutions. . . . We demand with unequivocal clarity that the Security Council fulfill its commitments toward Iraq . . . [and] fully and totally lift the blockade."[6]

That was a significant statement. It had been issued by Iraq's top leadership. It was unequivocal. And it suggested that Iraq was not going to give up any more of its proscribed weapons capabilities or to accept the status quo much longer.

The First Iraq Crisis

Richard Butler, a blunt-spoken Australian, replaced Ekeus on July 1. Even before Butler formally became UNSCOM chairman, he met with Tariq Aziz. Following their first meeting, on June 26, Butler asserted that he would work with "objectivity, including as far as possible avoiding political statements."[7] On his first trip to Baghdad, Butler hailed "a new spirit of cooperation between Iraq and UNSCOM and a new sense of determination to get done, as soon as possible, the job of eliminating Iraq's weapons of mass destruction."[8] Veteran Iraq watchers followed his statements with concern, because they seemed to suggest that Butler would be much less tough on Iraq than was his predecessor.

But it turned out merely that Butler was making a vigorous effort to give Iraq an opportunity to establish a new, more cooperative relationship with UNSCOM. And Iraq was not interested. Instead, it set the stage for confrontation. In mid-September, it blocked two inspections; two weeks later, it blocked three more.

When UNSCOM's next report appeared, it was indistinguishable from the reports issued under Ekeus.

So the crises began. The United States turned again to the Security Council. On October 23, the council issued UNSCR 1134. It repeated the language of the two previous resolutions and added that six months hence, if Baghdad were not cooperating with UNSCOM, travel sanctions would be applied.

Iraq's foreign minister, Muhammad al-Sahaf, then in New York, denounced the United States and Great Britain for imposing "their own sick motives and norms" on the council. And he asserted, "It is now up to our leadership and the Iraqi people to decide what are the alternatives to protect our country against this flagrant preparation to impose more injustices."[9]

And that is just what happened. The RCC and party leadership met the next day to discuss the "unfair resolution." They announced that they would "refer the matter to the National Assembly . . . to make the necessary recommendations."[10] The National Assembly met and issued a statement on October 27. It recommended that the government "make a decision on freezing relations with UNSCOM."[11]

Two days later, the Iraqi leadership issued another statement. It asserted that waiting for the Security Council to lift sanctions was "futile," because the United States had "evil intentions." Therefore, "We, as a first step, decide . . . not to deal with those carrying U.S. citizenship, who work [in] UNSCOM. . . . We will ask those Americans inside Iraq to leave Iraqi territory within one week of the date of informing the Security Council of our decision."[12]

The statement also explained that the leadership had conducted a "profound study" of meetings it had held on the issue, "particularly since March 16, 1997." That day, Iraq's foreign minister briefed the cabinet on discussions he had held with UNSCOM and the International Atomic Energy Agency. His report suggested, of course, that Iraq was not about to receive a clean bill of health on its weapons program. So Saddam began to prepare for what was to come. He waited until Butler had issued his first report, and then initiated the series of crises that would drive UNSCOM from Iraq.

Something similar had happened in 1994, when Iraq lunged at Kuwait. Then, Saddam had given his diplomats a chance to do what they could to get sanctions lifted. But he had had a deadline in mind, and had anticipated that they would fail. And when their time ran out, he was ready with an alternative, defiant, and forceful strategy.

After Baghdad's announcement that it would expel the American members of UNSCOM, the Clinton administration adopted the stance that it was a UN problem. On November 2, America's UN envoy, Bill Richardson, asserted: "This is not a fight between the United States and Iraq. This is Iraq's confronting the United Nations."[13] So UN Secretary General Kofi Annan dispatched a delegation to Baghdad on November 5. It left two days later, empty-handed. One member of the group observed that the Iraqis "seemed to have made up their mind even before we arrived."[14] Yet Baghdad also said that as a "concession" to Annan, it would postpone the implementation of its decision to expel the Americans. More negotiations followed.

The Clinton administration geared up to mobilize support for yet another UN resolution. "Top officials . . . fanned out across Washington's televised talk shows to present a firm, unified position in advance of the Security Council meeting," the *New York Times* reported.[15] And for the first time in two years, the administration publicly explained the dangers posed by Saddam's unconventional weapons programs, as they had become known after Kamil defected.

In a Veterans Day address, Clinton said, "I want every single American to understand what is at stake here—these [weapons] inspectors, since 1991, have discovered and destroyed more weapons of mass destruction potential than were destroyed in Iraq in the entire Gulf War."[16] Six days later, on November 17, he affirmed: "We must not allow the twenty-first century to go forward under a cloud of fear that terrorists, organized criminals, drug traffickers will terrorize people with chemical and biological weapons. . . . It is essential that those inspectors go back to work."[17]

Most memorably, Secretary of Defense William Cohen appeared on national television with a five-pound bag of sugar. To illustrate the dangers posed by Saddam's weapons, Cohen held up the bag and explained, "This amount of anthrax could be spread over a city—let's say the size of Washington. It would destroy at least half the population of that city."[18]

On November 12, the Security Council adopted another resolution (1137). It imposed travel sanctions on lower-ranking Iraqi officials. The next day, Iraq announced that it was immediately expelling the American staff of UNSCOM. Butler withdrew most UNSCOM personnel, leaving only a skeleton crew at UNSCOM's monitoring center.

Iraq's move seemed to catch the administration unprepared. Washington seemed to think that the travel sanctions would cause Baghdad to back down, but it did not. Clinton ordered a second aircraft carrier to the Persian Gulf, and speculation quickly turned to a possible U.S. strike on Iraq.

Madeleine Albright, who was then the secretary of state, was scheduled to attend an economic conference related to the Arab-Israeli peace process in Qatar on November 16, after which she was to travel to Pakistan. Instead, she made a hurried detour to Saudi Arabia on November 17. She met with Crown Prince Abdullah, who had assumed responsibility for the country's affairs two years earlier, when the king suffered a serious stroke. Any significant U.S. attack on Iraq would require the use of Saudi airspace, yet it was unclear if the Saudis were prepared to support a strike on Iraq.

But Saddam offered a way out. On November 16, he stated that he wanted to avoid a military confrontation and, "if a solution was made available through dialogue leading to the fulfillment of the UN Security Council's obligations toward Iraq, we would be happy."[19] The next day, an aide traveling with Albright suggested that the United States might be willing to offer Iraq "a little carrot." Among the U.S. proposals was one to increase by 50 percent the value of oil that Iraq was allowed to sell under the terms of UNSCR 986, which provided for the supervised sale of Iraqi oil for humanitarian purposes.

Serious negotiations had begun. On November 18 Aziz went to Moscow, where he met with Russian foreign minister Yevgeny Primakov. Two days later, Primakov summoned the Security Council's Permanent Five members (Perm-5)—the United States, Great Britain, France, Russia, and China—to a late-night meeting in Geneva. There, in the early hours of November 20, it was announced that UNSCOM's commissioners, a technical body consisting mostly of scientists, were to meet in an unprecedented emergency session to discuss "ways to make UNSCOM's work more effective."[20]

UNSCOM would return to Iraq. Later that day, Sandy Berger affirmed, "There are certainly no concessions, no deals that have been made."[21] But of course, a deal had been made. The integrity of UNSCOM's work had been impugned. Convening the commissioners to examine its work implied that UNSCOM was, at least in some respect, deficient. It might have seemed a small point; but, of course, it was not the end of the matter.

The Second Iraq Crisis

Iraq's National Assembly met in an emergency session on November 27. It issued a statement calling on UNSCOM to "expedite the closure of its files and end inspections in Iraq within a maximum period of six months."[22] It was not long before the next crisis began.

On December 15, Butler visited Baghdad. Iraqi officials told him that Saddam would not allow his palaces to be opened for inspections, but it was not clear what that meant. A month latter, UNSCOM inspector Scott Ritter was pursuing intelligence reports that Iraq had used political prisoners to test the lethality of its biological and chemical agents. On January 12, he led an inspection. At the end of that day, Baghdad, denouncing Ritter as a spy, declared that his team included too many Americans and could not continue its work. Iraq blocked Ritter's team the next day, and the Security Council issued a statement on January 14. On January 16, Great Britain announced that it was sending an aircraft carrier to the Gulf to back up the two American carriers in the region.

In his annual speech marking the start of the Gulf War on January 17, Saddam was defiant: "Iraq—people, leadership, and representative bodies, and on all levels—is irrevocably determined to wage the greater jihad for the lifting of the blockade. Unless the UN Security Council decides to fulfill its obligations toward Iraq . . . Iraq is determined to take a stand that conforms with the recommendations of the people's representatives in the National Assembly."[23] Saddam also urged Iraqis to prepare for jihad, and the Defense Ministry issued a call for the population to register for "popular training."

Butler arrived in Baghdad two days later. Aziz presented him with a list of eight areas—"presidential sites"—that were off-limits to UNSCOM. Aziz also suggested that there should be a moratorium on further talk about the inspection of "sensitive sites" until April. Butler had agreed to an Iraqi request to convene independent panels of experts to evaluate Iraq's "progress" in dismantling its proscribed weapons programs. UNSCOM believed that the independent reviews would confirm its position. But Aziz asserted that they would confirm Iraq's position and clear up all outstanding issues. Hence, there was no need to discuss the inspection of sensitive sites.

On January 23, Butler reported to the Security Council on his dismal trip. The administration again prepared to attack Iraq. "There'll be one final round of diplomacy, and then an ultimatum, and then we act," a National Security Council official explained.[24] But others questioned what the United States would achieve by a military strike, particularly as it would probably mean the end of UNSCOM. The administration had no good answer. Indeed, "a senior administration official reacted angrily to that argument. 'The question is not, What's next? after a strike, but What now?' "[25]

Yet administration officials began to modify their claims about what a military strike would achieve. On February 6, Clinton said that the aim of any strike would be to "substantially reduce or delay" Iraq's ability to develop and use unconventional weapons.[26] On February 17, he asserted: "If . . . we have to use force, our purpose is clear: We want to seriously diminish the threat posed by Iraq's weapons of mass destruction."[27]

But if the threat from Iraq could be represented by just five pounds of anthrax, how could a bombing campaign "seriously diminish" that threat? Didn't we have to get rid of Saddam? Senator John Kerry, a Democrat from Massachusetts, noted the "disconnect" between the threat and what the administration proposed to do. A Vietnam veteran, Kerry suggested that he would be ready to support even a ground war to eliminate Saddam. Advised that he was "way ahead" of dominant opinion in Washington, Kerry responded, "I believe in this."[28] Indeed, many in Congress began to call on the administration to develop a policy to overthrow Saddam. Generally, they urged the administration to return to the option it had so carelessly thrown away the year before, when it had allowed Saddam to attack the Iraqi National Congress.

On February 18, a Town Hall meeting was held at the University of Ohio. The president's most senior foreign policy advisers—Berger, Albright, and Cohen—were to explain their position on Iraq. The meeting was broadcast live on CNN to the whole world, including Baghdad. But the audience's response was totally unanticipated. It repeatedly heckled and interrupted the speakers. And it asked tough and pointed questions.

An elderly man explained, "I spent twenty years in the military. My oldest son spent twenty-five. My youngest son died in Vietnam. Six months later, his first cousin died in Vietnam. . . . If push comes to shove and Saddam will not back down . . . are we willing to send troops in and finish this job?"

An angry young man protested, "We are not going to be able to stop Saddam Hussein, we are not going to be able to eliminate his weapons of mass destruction. . . . President Clinton admitted it. All he wants to do, Clinton said, was send a message to Saddam Hussein. If he wants to send a message, we the people of Columbus and central Ohio and all over America will not send messages with the blood of Iraqi men, women, and children. If we want to deal with Saddam, we deal with Saddam, not the Iraqi people."[29]

Almost no one outside the administration, including the audience in Ohio, understood that the White House was then looking for a way out of the crisis that did *not* involve attacking Iraq. In the last week of January, the administration had worked out "Pol-Mil Plan Iraq" (pol-mil indicates "political-military"), with "timelines" for successive stages of "rhetorical escalation and military deployment."[30] The original idea was to threaten Saddam with military force, in the expectation that he would back down. But, if not, the United States would strike. As late as February 11, during a meeting of the Perm-5, Richardson dismissed any notion that Annan might go to Baghdad for last-ditch negotiations. Privately, Richardson told Annan, "You can't go. You can't box us in."[31]

By the next day, however, the president had changed his mind. On February 15, Albright quietly flew to New York to meet with Annan. She told him to disregard what Richardson had said. She also explained the administration's "red lines" (the crossing of which would not be tolerated). UN inspectors had to have full access, and operational control of inspections had to remain in UNSCOM's hands. Any agreement Annan reached with Baghdad had to be in writing.[32]

On February 19, Annan left for Iraq. To smooth his way, the Security Council passed a resolution the next day more than doubling the value of oil that Iraq could sell under UNSCR 986. From $2 billion every six months, the sum was raised to $5.6 billion. Henceforth, Iraq could sell all the oil it could produce.

The Security Council also worked out what was envisaged as a face-saving compromise. Diplomats would accompany UNSCOM teams when they inspected presidential sites. That was intended as a token of respect for Iraq's sovereignty and dignity. But surprise was essential to a successful inspection. The "diplomatic nannies," as former weapons inspector David Kay dubbed them, would ensure that Iraq would have ample warning of such inspections.

On February 23, Aziz and Annan reached an agreement and signed it. Upon returning to New York, Annan described Saddam as someone "I can do business with. . . . He was serious when he took the engagement. I think he realizes what it means for his people."[33] And Clinton welcomed the agreement: "The government of Iraq has made a written commitment to provide immediate, unrestricted, unconditional access for the UNSCOM weapons inspectors. . . . [But] what really matters is Iraq's compliance, not its stated commitments. . . . In the days and weeks ahead, UNSCOM must test and verify."[34] Many in Congress were nevertheless outraged. Senate Majority Leader Trent Lott denounced the agreement as "appeasement," charging that "it is always possible to get a deal if you give enough away."[35]

UNSCOM regarded the agreement similarly. Some on the staff even considered attending Annan's triumphal UN reception with black umbrellas, evoking Neville Chamberlain's 1938 agreement with Hitler. In fact, Butler was discussing Annan's agreement with close aides when Albright telephoned. After the conversation ended, he explained, "Madeleine is very nervous. The heat coming down from Trent Lott is upsetting the administration. She wants me to hold a press conference and endorse the Kofi Annan agreement."[36]

Of course, the new agreement was most unlikely to end the Iraq crises, as the administration recognized. "What strikes many of the president's advisers," the *Washington Post* reported, "is that the next round . . . will begin exactly where the last one left off: another probe, another threat, another struggle to channel domestic and international support for military force. 'God knows what's going to happen,' said a high-ranking administration official."[37]

And what had been accomplished? The U.S. military build-up had cost more than $600 million. It had roiled the Middle East. In the West Bank and Gaza, Palestinians demonstrated in support of Saddam, even calling on Iraq to hit Israel with missiles and chemical weapons.[38] The head of Egypt's Muslim Brotherhood warned of "dire consequences" if the United States were to strike Iraq, and pro-Iraqi demonstrations were held throughout the region. And the "World Islamic Front" suddenly emerged. It consisted of a handful of Islamic extremists based in Afghanistan, including Usama bin Ladin, as discussed in chapter 18. On February 23, 1998, it issued its first statement, denouncing the "great devastation inflicted on the Iraqi people" and calling on all Muslims "to kill the Americans and their allies."[39]

Interlude

As UNSCOM resumed its work in Iraq, U.S. officials repeatedly affirmed that Baghdad would cooperate fully with UNSCOM, as it had committed to do, or face serious consequences. They said so to the American public, the U.S. Congress, and America's allies. Thus, Sandy Berger affirmed, "If Iraq follows through on its commitment, the inspectors will for the first time have unrestricted, unconditional access to all suspect sites—including sites the Iraqi government previously had declared off limits. If Iraq reneges, we will respond powerfully."[40] Three days later, Albright told Congress: "We are continuing our effort—through diplomacy backed by the threat of force—to see that Iraq complies. . . . If Iraq violates the agreement, there will be . . . a forceful response."[41] Subsequently Albright toured Europe and Canada, reiterating that position.

Two U.S. aircraft carriers remained in the Persian Gulf, as did a British carrier. To all appearances, it looked as if the United States had succeeded in imposing its will on Iraq. Baghdad, it seemed, was lying low, cooperating with UNSCOM to avoid a U.S.-British military strike.

But the situation was more complicated. The United States was also trying to avoid a confrontation with Baghdad. UNSCOM returned to Iraq in early March, and Ritter was to lead the inspections. The inspectors assembled in Bahrain on March 3 to prepare for their work. Upon his arrival there, Ritter was given a message to call Butler. Butler told him, "Madeleine is not happy about your being chief inspector. . . . We need to consider options."[42]

Bill Richardson, whom Ritter regarded as the only "stand-up person" on Clinton's national security team, learned from his political counselor on UNSCOM affairs that Butler was about to replace Ritter. Richardon's reaction is recounted below:

> "What the hell is Butler up to?" Richardson asked. "I thought he had committed to Ritter staying on." The counselor replied that . . . Albright had put pressure on Butler. Richardson was incensed. He told the counselor to get Butler on the phone. . . . [B]ut Butler was nowhere to be found. . . .
>
> Richardson held up his hand, signaling silence. It was Madeleine Albright on the line. "Listen, Madeleine," Richardson said. "Butler is really irritated at your intervention. He says Ritter's the guy and that he resents your meddling into his affairs. He may resign."
>
> Albright was flabbergasted. "But I just talked to him. . . . He didn't indicate any of this when I was on the line. I don't believe it."
>
> Richardson pressed, "Believe it," he said. "Ritter's on, or Butler's out." Still stunned, Albright said she would call Butler right away to clarify the matter.
>
> Richardson hung up and looked at the counselor, who was standing by in disbelief. "This is high stakes, my friend," Richardson said with a laugh. "Find Butler before Madeleine does or we're both screwed."[43]

Richardson's aide succeeded in reaching Butler, and Ritter led the inspections.

After weapons inspections resumed, the administration did urge Butler to search a high-profile site, and the United States was prepared to attack Iraq if it failed to cooperate. That plan reversed the decision taken on February 12 to seek a face-saving way out of striking Iraq. The reversal was probably prompted by the strong congressional criticism of the Aziz-Annan accord.[44] The administration suggested that UNSCOM search Iraq's Defense Ministry. The search of a site that big, however, required considerable preparation. The annual Muslim pilgrimage to Mecca, the Hajj, was to start on March 15. Any strike would have to end by then, and the U.S. military had plans for a seven-day bombing campaign. So the Defense Ministry had to be inspected by March 8. The inspection was indeed carried out that day. It did not much matter that UNSCOM did not really have enough time to plan for it; the building had been emptied of its contents.[45]

Inspections of "presidential sites," replete with diplomatic nannies, were carried out between March 26 and April 3. Those inspections proved equally fruitless. The sites were sanitized, and Butler later described the inspections as "ludicrous."[46] Nonetheless, administration officials repeatedly asserted that their policy had succeeded. UNSCOM was visiting places it had never been before, and Iraq was cooperating.

Of course, all that exploded in August, when Ritter resigned, protesting that senior U.S. officials were actually blocking UNSCOM inspections. At first, the administration denied the charge. Albright stated at an August 14 press conference, "Let me make this perfectly clear. I do not tell Chairman Butler what to do." But she had.[47] And the editors of the *New York Times* denounced the "illusory inspections," as did many others.[48] An outraged Congress demanded answers, but the most senior administration officials balked at appearing before Congress. The job fell to Assistant Secretary of State for Near Eastern Affairs Martin Indyk. Indyk told the Senate, "The United States has never impinged on UNSCOM's integrity or attempted to dictate its decisions." When pressed, Indyk acknowledged that the United States had "asked questions" about some inspections, and they had been "adjusted."[49] Former UN ambassador Jeane Kirkpatrick, testifying after Indyk, explained that although she had listened sympathetically to what he said, she could not understand U.S. policy. She also noted that it was "unacceptable" for the administration to give the American public the impression that it was vigorously pressing a policy of inspections and then not pursue such a policy.[50]

Behind the scenes, the administration had, in fact, been developing a new policy on Iraq that was quite different from its declared policy. *Washington Post* columnist Jim Hoagland was the first to catch wind of it. In late April, Hoagland wrote that the administration would no longer support UNSCOM with the threat of force. Rather, it would threaten force only if a clearly dangerous situation emerged—if Iraq mobilized troops against Kuwait or Saudi Arabia, or if Iraq openly deployed chemical or biological weapons.[51] The new policy was given the name "deterrence," as opposed to the previous policy of "containment." Under Secretary of State Thomas Pickering appeared before the Senate on May 21. When asked about the press reports suggesting that the administration was developing a new, less vigorous posture on Iraq, he said they were not accurate.[52]

Regrettably, the press had it right. In early May, it was reported that the administration was considering the reduction of the number of U.S. aircraft carriers in the region.[53] Yet the matter remained ambiguous. As late as Friday, May 22, on the eve of the Memorial Day weekend, the White House denied that a decision had been made; but only two days later, the Pentagon announced that one of the two carriers stationed in the Persian Gulf was on its way out. The U.S. military presence had been halved with virtually no public discussion. A new policy was in place.

But Saddam was also moving ahead with his own plans. Following the end of the second Iraq crisis, the Security Council began to conduct sanctions reviews. On April 17, UNSCOM issued its biannual report, explaining that it had made virtually no progress. On April 27, the council reviewed sanctions and kept them in place.

Four days later, on May 1, 1998, the Revolutionary Command Council and Ba'th Party leadership met. They issued a strong statement, addressed to the Security Council. It began, "Wars in general happen primarily because of a maximum feeling of injustice, making war the only means to remove the injustice." And it warned, "The inability of the Iraqi people to see a lifting of sanctions after eight years . . . will lead to dire consequences."[54]

Iraq reiterated that statement four times over the next three months. On August 3, Butler arrived in Baghdad for more talks. But Iraq was not interested in further discussions. Aziz demanded that Butler declare that Iraq no longer possessed any proscribed weapons. Of course, Butler was not prepared to do so, and he left Baghdad the next day.

On August 5, 1998, Iraq announced "Suspension Day." The National Assembly met that morning and recommended breaking ties with UNSCOM. The Iraqi leadership then issued an angry statement, asserting that Iraq had "fulfilled all its obligations with the objective of lifting the blockade." And it announced that while Baghdad would allow UNSCOM to continue its monitoring activities, it was suspending weapons inspections until its demands were met. Two days later, the U.S. embassies in Kenya and Tanzania were bombed simultaneously, as is discussed in chapter 18.

The Third Iraq Crisis

Although Iraq's "suspension" of inspections was a far more serious challenge to UNSCOM than anything Iraq had done before, the administration's response was muted. The new policy of "deterrence" was in effect. It was left to the Security Council to find some way of persuading Iraq to allow UNSCOM inspections to resume. And the Security Council moved tortuously. On September 9, a month after inspections had ceased, the council passed yet another resolution demanding that Iraq cooperate with UNSCOM. UNSCR 1194 suspended the sixty-day sanctions reviews until Iraq renewed its cooperation with UNSCOM, and it also offered to conduct a "comprehensive review" of the extent to which Iraq's unconventional weapons programs had been destroyed, once Baghdad allowed the inspections to resume. The measures were meant as a stick and a carrot, but both were limp.

And Congress was growing increasingly frustrated. In early October, by overwhelming majorities in both houses, it passed the Iraq Liberation Act. The measure authorized the provision of $97 million in U.S. military equipment to support an insurgency against Saddam. The Clinton administration let it be known that it intended to veto the legislation.

The Security Council continued its work on the "comprehensive review." On October 30, nearly three months after inspections had ceased, the council finally reached agreement on the guidelines for such a review. Baghdad responded immediately. The next day it announced that it would terminate UNSCOM monitoring as well.

Again, Saddam's defiance seemed to catch the administration by surprise. William Cohen cut short an Asian tour and hurried back to Washington. On November 5, the Security Council passed yet another resolution demanding that Baghdad cooperate with UNSCOM. And the United States again geared up to strike Iraq.

On Veterans Day 1998, Clinton warned (as he had the year before) about the dangers posed by Saddam's weapons: "If the inspectors are not permitted to visit suspect sites or monitor compliance at known production facilities, they may as well be in Baltimore, not Baghdad. That would open a window of opportunity for Iraq to rebuild its arsenal of weapons and delivery systems in months—I say again, in months—not years. A failure to respond could embolden Saddam to act recklessly."[55]

That day, November 11, UNSCOM withdrew its remaining staff from Iraq. Two days later, on the eve of the planned U.S. strike, Sandy Berger met with Trent Lott. Berger told Lott that the White House would adopt a new policy. It would bomb Iraq and then adopt the Iraq Liberation Act to get rid of Saddam.

But that night there was a flurry of activity at the United Nations. Kofi Annan issued a last-ditch appeal to Baghdad. That elicited a faxed letter from Tariq Aziz, saying that UNSCOM could return. Berger heard about the letter on CNN on the morning of November 14, but he had not seen it. U.S. planes were already in the air, ready to start bombing at 9:00 A.M. Fifteen minutes before the attack was to begin, on Berger's advice, Clinton ordered a "pause" in the military operation.[56]

When the White House finally saw Aziz's letter, it proved unacceptable. It seemed to tie UNSCOM's return to a review of sanctions that would lead to their quick removal. So Iraq sent another draft, which was also unacceptable. Iraq sent a third, which finally passed muster. But at that point, how could any reasonable person have confidence in an Iraqi promise—however worded—to cooperate with UNSCOM?

Nevertheless, when Clinton addressed the nation the next day, November 15, he presented the letter as a resolution to the crisis. He asserted that Iraq had backed down. It had "committed to unconditional compliance." That, Clinton affirmed, was "the outcome we preferred," as the weapons inspectors "have been and they remain the most effective tool to uncover, destroy, and prevent Iraq from rebuilding its weapons of mass destruction." Clinton stressed that the United States would "remain vigilant; we will keep up the pressure; we will be ready to act." And he listed five demands Iraq had to meet. They included providing UNSCOM "unfettered access" and turning over all documents that UNSCOM requested. Clinton also introduced a new and more forceful element into U.S. policy. He explained that "over the long term," the best way to address the threat from Saddam was to overthrow him, and he vowed to work with Congress and the Iraqi opponents of Saddam to implement the Iraq Liberation Act.[57]

The Fourth Iraq Crisis

UNSCOM resumed its work in Iraq on November 17. As part of the understanding that led to UNSCOM's return, Richard Butler was to report on Iraq's cooperation with UNSCOM. If the report were favorable, there would be a "comprehensive review" of Iraq's compliance with the UN resolutions. Butler estimated that it would take four weeks to prepare the report.[58]

But already on November 23, Butler informed the Security Council that he had given Iraq a list of needed documents and Baghdad had refused to provide them. There was little U.S. response. The editors of the *Washington Post* chided,

> [You] may recall, because it happened only one week ago, that Iraq promised full cooperation with U.N. arms inspectors. You may recall that President Clinton proclaimed this promise a major accomplishment. . . . You may recall also that administration officials warned that any Iraqi failure to provide requested documents or otherwise cooperate with U.N. inspectors would result in bombing without further warnings. Well, Iraq has

now refused to provide such documents. And the American response? More warnings—and pretty unconvincing ones at that.[59]

Other problems ensued. On November 23, an UNSCOM helicopter was buzzed by an Iraqi helicopter. On November 26, UNSCOM was denied access to a site. On December 4, it was denied access to another.

But still it appeared that the United States would not act, or at least not immediately. The Muslim fasting month of Ramadan would begin on December 18. The *Washington Post* (on December 7) reported, "diplomats and Clinton administration officials said there is no prospect of enforcing cooperation now with . . . Ramadan about to begin."[60] According to a *New York Times* report, the administration was, in fact, so far from pursuing military action that it was prepared to agree to the "comprehensive review" demanded by Iraq, even if Butler reported that Iraq was not cooperating fully. The rationale was that any such review would highlight the extent of Iraq's noncompliance and thus would "backfire" for Iraq.[61]

But in Washington another, quite different issue was emerging. On December 11, the House Judiciary Committee voted to approve three articles of impeachment against the president. On December 12, it voted to approve a fourth. The full house was to vote on the matter on December 17.

On December 15, Butler submitted his report to the Security Council. Iraq had provided some documents, but not others. It had allowed most inspections to proceed, but had blocked four. Butler concluded, "Iraq did not provide the full cooperation it promised."[62] Even so, UNSCOM's situation was better than it had been at any time since August, when inspections had ceased.

Nevertheless, the next night, the United States and Great Britain began a military attack on Iraq—Operation Desert Fox. As the bombing started, Clinton addressed the nation: "Over the past three weeks, the UN weapons inspectors have carried out their plan for testing Iraq's cooperation. . . . Last night UNSCOM's chairman, Richard Butler, reported the results to UN Secretary General Annan. The conclusions are stark, sobering, and profoundly disturbing," Clinton announced. "Without a strong inspection system, Iraq would be free to regain and again to rebuild its chemical, biological, and nuclear weapons programs. . . . Mark my words, [Saddam] will develop weapons of mass destruction. He will deploy them and he will use them." Clinton also said, as he had the month before, that the "best way" to deal with the Iraqi threat was to establish a new government in Baghdad, and he vowed to "strengthen our engagement with the full range of Iraqi opposition forces."[63]

Senior administration officials repeatedly stated that the start of Ramadan would not dictate the schedule of military operations. "We will take whatever time is necessary to carry out the operation," asserted Cohen.[64] "We are sensitive to the beginnings of Ramadan, but this campaign will continue to its completion," said Albright.[65]

Yet on the evening of December 19, after four nights of strikes, Clinton announced that the bombing campaign was over: "Our objectives in this military action were clear: to degrade Saddam's weapons of mass destruction. . . . I am confident we have achieved our mission."

Others disagreed. Former secretary of state Henry Kissinger said, "I would be amazed if a three [sic]-day bombing campaign made a decisive difference. . . . I think we lost the opportunity we had three times during the year to do something decisive, or at least to try to do something decisive."[66] The editors of the *New Republic* gave Clinton the benefit of the doubt. Although questioning his stated motives—they pronounced as "disingenuous" the claim that Butler's report, rather than the impending impeachment vote, had triggered the strike—they believed that Clinton had done the right thing, and they hoped it would mark the start of a more aggressive posture toward Saddam.[67] A. M. Rosenthal of the *New York Times,* however, believed that little would come of the strike and little would follow it, and he called on Clinton to resign.[68]

The American public, on the whole, rejected the notion that the U.S. strike on Iraq was precipitated by the impending impeachment vote. As the new year began and the Senate started to hear the charges against the president, the majority of Americans opposed his removal from office. To be sure, most said (according to opinion polls), Clinton had lied; but it was just about sex.

UNSCOM left Iraq as Operation Desert Fox began and never returned. And despite the administration's claim to have a new policy to overthrow Saddam, it did little to follow through.

It is disturbing to consider that senior U.S. officials, including the president, could have described UNSCOM's importance—and the dangers Saddam would pose in its absence—in the terms that they did, and then do virtually nothing about the problem once the inspectors left Iraq. I asked several experienced people in Washington who had direct dealings with the administration over Iraq, how was this possible? Time and again the answer was, "the news cycle." One former Clinton administration official explained to me that the administration remained stuck in the habits and style of a political campaign. Whatever the problem, a very high priority is placed on getting positive news coverage—spin. If you succeed in producing good news the first day, that will help the next day, and so on. Strategic thinking, while welcome, is "extracurricular," this former official explained. As Sandy Berger himself affirmed to the *New York Times*, "he prefers to 'worry about today today and tomorrow tomorrow.' "[69]

Yet the world is not born anew every morning. Today is the product of a thousand yesterdays. If there is no constancy in dealing with a serious problem like Iraq, if it is consistently redefined according to the circumstances, don't all yesterdays' spins constrain tomorrow's options? And all that spinning might, in time, result in a potentially dangerous situation in which almost no one can quite comprehend the threats facing the country or how they came about, let alone how to address them.

13

Other Links

I t is doubtful that all the conspirators involved in the World Trade Center bombing have been identified, let alone brought to justice. Indeed, that was Jim Fox's view.[1] Particularly if the Trade Center bombing were a foreign intelligence operation, there would have been individuals indirectly involved, providing logistical support and otherwise facilitating the terrorists' work. Those supporters would have done little for which they could be held criminally responsible (recall the structure of the Abu Nidal cell in London, described in chapter 2). Treating the case as a judicial matter significantly increases the likelihood that the investigation will overlook significant links.

Telephone records are especially useful for the investigation of a complex conspiracy. A conspiracy entails communications, and the telephone is one of the most common means of communication. The telephone records can reveal the existence of conspirators whose role in a terrorist plot may have been indirect and less readily apparent than that of individuals involved in activities such as building a bomb or acquiring components for it.

At the time of the bombing, Abdul Rahman Yasin's brother Musab was a thirty-three-year-old Ph.D. student in electrical engineering at the City University of New York. He also taught electrical engineering at Hudson County Community College.

Until at least mid-May 1992, Musab Yasin's telephone was listed with the telephone company under his own name, as indicated by the Trade Center bombing defendants' phone records.[2] But by the end of August, Yasin had a new and unlisted phone number, registered with the telephone company under a false name, "Josie Hadas." Thus, someone trying to locate Yasin through phone company records would not have been told that such an individual had an unlisted number in Jersey City. He would have been told that the phone company no longer had any record of Yasin in the area. Also, Yasin's phone bill was not sent to his apartment at 34 Kensington Avenue, but was sent to an address in Brooklyn.[3] As one New York newspaper, citing law enforcement agents, observed, the residents of Yasin's apartment "took extraordinary pains to maintain anonymity."[4]

In June 1993, Yasin filed a sworn affidavit, under penalty of perjury, claiming that he had moved into 34 Kensington Avenue "in or about September 1992," at the same time that Salameh (who already lived in the building) moved out of what would become Yasin's apartment, number 4, and into number 8. But Yasin moved into 34 Kensington significantly earlier than September 1992. His name, address, and phone number appear in the 1992 Jersey City telephone book (although not the 1991 phone book), and the phone book was published in January, as is indicated on its cover. Thus, Yasin was living at 34 Kensington already in January 1992. Why would he misstate the date on which he moved into his residence?

One possibility is that he simply made a mistake in his affidavit. His memory was grossly inaccurate. Or there was a typing error in his statement; perhaps he moved into the apartment in September 1991. The other possibility is that Yasin deliberately misstated when he moved into his apartment. By claiming that he moved into 34 Kensington in September 1992, he blurred the fact that prior to the start of the Trade Center bombing conspiracy, his phone was registered in an ordinary fashion. But once the plot got underway—and his brother is an indicted fugitive in the Trade Center bombing who stayed in his apartment—the listing of his telephone in phone company records became quite secretive.

The building in which Yasin lived, 34 Kensington Avenue, was so central to the plot that one law enforcement official dubbed it "the conspirators' den."[5] When Salameh rented the van to carry the bomb, he listed Yasin's phone number on the rental form.[6] When Ramzi Yousef and Salameh were injured in a one-car accident on January 24, 1993, they gave hospital officials their address as 34 Kensington Avenue, apartment 4.[7] Probably, they did not want to reveal the existence of the Pamrapo Avenue apartment where they actually lived and where they were making the bomb.

It was widely reported that Salameh, a notoriously bad driver, was behind the wheel at the time of the accident, but Yousef told the examining doctor otherwise. The doctor's report on Yousef begins, "Patient is a twenty-five-year-old male, involved in a motor vehicle accident. Patient was driver."[8] Yousef had already managed to postpone one asylum hearing, because he did not have a lawyer. Another was set for January 26. Yousef remained in the hospital until January 29, thereby succeeding in postponing his second scheduled asylum hearing.[9]

Thus, the accident was very convenient, although it totaled Salameh's car, which was abandoned as undrivable. But another car was available, registered in the names of Ramzi Yousef and Muhammad Salameh. It belonged to Yasin and was registered to the same Brooklyn address to which his phone bills were sent.[10]

Jim Fox, who saw the bombing as an Iraqi intelligence operation, was very suspicious of Yasin, as he once told me. But others, including New Jersey FBI, were not. In fact, New Jersey FBI considered Yasin to be cooperative and a useful source of information. He was never charged with any involvement in the bombing.

"Waly Samar"

On February 2, 1993, a new number appeared on the Trade Center bombing telephone records. The phone belonged to a dormitory room at New York's Hunter College and it was listed under the name "Waly Samar." It appeared only on Yasin's records, and it was called frequently from Yasin's apartment throughout the month of February and into early March.

Most notably, on February 28, two days after the Trade Center bombing, a call was placed from "Waly Samar's" dormitory room to a number in Islamabad and billed to Yasin. The call was to the Saudi Red Crescent, and it lasted ninety-five minutes. It was made at 12:45 P.M. local time, when it would have been 10:45 P.M. in Pakistan—not an hour when people are usually in an office. That is a sign of intelligence "tradecraft." One cannot tell who took that call, just as one cannot tell who made it.

Long phone calls are significant, as a deputy to the defense attaché at the Israeli Embassy in Washington advised me. He looked at these phone records and pointed out the ninety-five-minute call. He suggested that it was likely to have been a reporting call about the Trade Center bombing.

In the days after the bombing, "Waly Samar's" dormitory room was practically the only toll number called from the Yasin apartment (for calls billed to the Yasin apartment, see chart 13–1). As the phone records indicate:

- No toll calls were made from the Yasin apartment on February 26, the day of the bombing.
- On February 27, Yasin's apartment called Mahmud Abu Halima once and "Waly Samar" twice. Several overseas calls were placed from "Waly Samar's" dormitory room and charged to Yasin's number, including calls to two numbers in Baghdad and one in Islamabad—the office of the Saudi Red Crescent.
- On February 28, the Yasin apartment made one call to Nidal Ayyad and one to an apartment unrelated to the bombing. A long call—ninety-five minutes—to the same Islamabad number, the Saudi Red Crescent, was placed from "Waly Samar's" dormitory room and charged to Yasin's number.
- On March 1, the Yasin apartment called "Waly Samar" three times, the only toll calls that day.
- On March 2, the Yasin apartment called "Waly Samar" two times, the only toll calls that day.
- On March 3, the Yasin apartment called "Waly Samar" three times, the only toll calls that day.

On the morning of March 4, Salameh was arrested. His arrest was announced on national television at the end of a noontime press conference. At 12:36 P.M., the first of four calls was made that day from Yasin's apartment to "Waly Samar."

Around 2:00 P.M., police sealed off the block in preparation for a search of 34 Kensington Avenue. The Yasin apartment made its last toll call that afternoon at 2:46 P.M.—to "Waly Samar." The occupants, including Abdul Rahman and his mother, were taken away by the FBI for questioning. Musab was at school. When he arrived home, he found FBI agents searching his apartment and he, too, was taken to the FBI office in Newark.[11]

The telephone records clearly suggest that the Hunter College dormitory room is of special interest. Who occupied that room and made the very long call to Islamabad that seems to have been a reporting call? Why did the Yasin apartment call that dormitory room so often in the days after the bombing?

The answers to these questions are not known. But there certainly appears to be at least one individual who was involved in the plot who has not been identified. Moreover, not everyone can rent a room at the Hunter College dormitory. Only full-time students at Hunter College can do so.

An individual who was a student at Hunter College appears with some frequency on the conspirators' telephone records, and he did some suspicious things with his own phone as well. Was that student involved in renting that room? If so, did he know to what purpose it would be put? That is a particularly relevant question because of his field of expertise: he is a microbiologist. An individual with a knowledge of microbiology would be particularly useful for any party that sought to carry out a biological terrorist attack.

In hopes of stimulating a more thorough investigation, I gave my analysis of the Trade Center bombing telephone records to an official at the New York district attorney's office. The official was impressed by the telephone analysis and did investigate the Hunter College student. But he came to the conclusion that he could do nothing; he pronounced it "frustrating."[12]

The Airline Pilots

In early 1995, Ramzi Yousef was arrested as a result of a failed attempt to bomb twelve U.S. airplanes, as discussed in chapter 15. Yousef was based in the Philippines at that time, and on the night of January 6 he accidentally started a fire while mixing explosives. For a month Yousef succeeded in eluding authorities,

CHART 13–1
LOG OF TELEPHONE CALLS BILLED TO THE
YASIN APARTMENT AFTER THE BOMBING

February 26: Friday
No calls

February 27: Saturday
10:11 A.M. Mahmud Abu Halima, 1 min.
11:05   Waly Samar, 5 min.
11:18   From Waly Samar to 964 1 556 1391, Baghdad, Iraq, 5 min.
11:24   From Waly Samar to 44 532 450 210, Leeds, U.K., 1 min.
11:34   From Waly Samar to 92 51 854 700, Islamabad, Pakistan, 36 min.
12:13 P.M. Waly Samar, 2 min.
12:34   From Waly Samar to 964 1 556 1391, Baghdad, Iraq, 45 min.
 3:31   From Waly Samar to 964 1 443 8091, Baghdad, Iraq, 24 min.
 4:45   Waly Samar, 4 min.

February 28: Sunday
12:08 P.M. Nidal Ayyad, 3 min.
12:46   From Waly Samar to 92 51 854 700 Islamabad, Pakistan, 95 min.
 8:49   Zainab Mahmod, Brooklyn, N.Y., 2 min.

March 1: Monday
11:51 A.M. Waly Samar, 1 min.
12:24 P.M. Waly Samar, 1 min.
 1:00   Waly Samar, 4 min.

March 2: Tuesday
 5:20 P.M. Waly Samar, 1 min.
10:55   Waly Samar, 6 min.

March 3: Wednesday
11:17 A.M. Waly Samar, 1 min.
 1:17 P.M. Waly Samar, 1 min.
 5:15   Waly Samar, 1 min.

March 4: Thursday
12:36 P.M. Waly Samar, 1 min.
12:59   Waly Samar, 1 min.
 1:59   Waly Samar, 1 min.
 2:46   Waly Samar, 1 min.

10:22	Shari Bagrian, 1 min.
10:22	Sharareh Bagherian, 1 min.
10:23	Sharareh Bagherian, 4 min.
10:27	Z.S. Abuasi, Rutherford, N.J., 2 min.

NOTE: Calls listed as originating "from Waly Samar" were made from his dormitory room and charged to the Yasin apartment.
SOURCE: Government Exhibit 818, *United States v. Muhammad Salameh et al.*

but a co-conspirator, Abdul Hakim Murad, was arrested that night.

As with Yousef, we do not know much about Murad. Murad's identification—like Yousef's—is based on a Pakistani passport stating his country of origin as Kuwait. Murad told authorities that he had grown up in Kuwait, where his father worked as an engineer for the Kuwait Petroleum Company, and that he knew Yousef from their days in Kuwait.[13] That claim remains uncorroborated by any independent source.

We do know that Murad, like Yousef, is Baluch. He was delighted when his lawyer showed him the entry on Baluchistan in the *Encyclopedia Islamica.* When the judge in their trial repeatedly offered them an Arabic translator (even though Yousef and Murad both speak flawless English, with an accent reminiscent of the colonial regions of the British Empire), Murad quipped, "Why not a Baluch translator?"

One point is particularly relevant to the possibility of unconventional terrorism in the planning of those behind Yousef: Murad is an airplane pilot. He obtained a pilot's license in the United Arab Emirates (UAE) in 1991 and came to the United States the next year. He attended several pilot training schools and received his U.S. commercial pilot's license on June 8, 1992. It came out during his trial, moreover, that he had a "friend" who studied for a U.S. pilot's license at the same time. Thus, those associated with the conspirators may include at least one person with a U.S. pilot's license: Murad's unnamed friend.

At the time of the Philippines bombing conspiracy, Murad reported that he had lost his pilot-license documentation. The Federal Aviation Administration sent him a replacement certificate at his UAE postal box on January 12, 1995, just after his untimely arrest in Manila.[14] Who retrieved this document from the UAE post office box? That pilot's license may well be in the hands of terrorists associated with Murad and Yousef.

The involvement of one or more airplane pilots in the conspirators' circle is of potentially great significance. As discussed in chapter 11, the most effective means of delivering a biological agent like anthrax is by an airplane to which is attached a spraying device that disseminates a cloud of the agent, so that it is carried over a populated area.

The Mystery Man

On the trip from Islamabad to New York after his arrest, Yousef talked with the FBI agents who accompanied him. He told them they could not write anything down, in the belief that if his statement was not written down, it could not be used against him in the trial. The agents nonetheless managed to produce a written account of what he told them.[15]

Yousef revealed, among other things, that there had been another participant involved in the bombing, someone who had driven the van that carried the bomb and who had left New York the same night as Yousef. He expressed surprise that authorities had not caught him. It was odd that Yousef gave out so much information about another conspirator. Did Yousef actually want the FBI to get him? Yousef could not know that an indictment had already been issued for Eyad Ismail, as it was still "sealed"—that is, secret—at the time of Yousef's arrest.

Eyad Ismail was the mysterious man known to Nidal Ayyad as "Rashid's friend." Ayyad had told authorities that Salameh had taken two men to JFK Airport on the night of the bombing—Yousef, whom he had known as Rashid, and another man, whom he had known as Rashid's friend (discussed in chapter 8). Ismail is not, as investigators had expected, a mysterious figure who supervised the bombing in its last stage. Ismail is a Palestinian who was twenty-one years old at the time of the Trade Center bombing. Following the bombing, he returned directly to his parents' home in Jordan, where authorities arrested him two and a half years later.

When Ismail was first indicted, on September 12, 1994, an extradition treaty did not exist between the United States and Jordan. Almost a year later, after a treaty had been concluded, Ismail was arrested, on July 30, 1995. Jordanian authorities questioned him for four days and then sent him to the United States.

After Ismail's arrest, Jim Fox remarked to me, "It looks like the FBI finally did what you did with the telephone records."[16] Indeed, all that was necessary to discover a connection between Ismail and the conspirators was to check Yousef's phone records against the passenger manifests of airplanes leaving JFK the night of the bombing. Ismail was on both—and, almost certainly, he was meant to be arrested.

What was Ismail's role in the conspiracy? Ismail's father had been a lathe operator at the machine shop of the Kuwait National Petroleum Company. Ismail (along with his older brother and six younger siblings) was born in Kuwait, where they were all raised. Ismail attended the Fahaheel secondary school in Kuwait for two years, from the fall of 1985 until the spring of 1987. He then finished his last two years of high school in Jordan. In October 1989 Ismail came to the United States as a student, enrolling in an intensive English language program at Wichita State University in Kansas. While Ismail's father still worked in Kuwait he was able to support Ismail financially, but after the Iraqi invasion the entire family left for Jordan. The PLO supported Saddam at that time, as did many Palestinians living in Kuwait (although there is no evidence that Ismail or anyone in his family did so). After the war, Kuwait did not generally allow Palestinians to return. Ismail's father thus lost his job, and in Jordan it was impossible for him to earn what he had been making in Kuwait.

Without his father's support, Ismail could not pay his tuition and he dropped out of school. In 1991 he left Kansas for New York, where he had an uncle and a cousin, both cab drivers. After a brief stint driving a taxi, he moved later that year to Dallas, to work at a restaurant owned by a family friend.[17]

Ismail's bosses and acquaintances recalled him as "not particularly bright."[18] Nor did he have any strong political or religious feelings. He liked life in America. He liked the freedom, the hamburgers, and the cowboy movies, and he was briefly married to a "tall blonde American beauty," as he described her to his family. But his family did not like his idling away in a distant, foreign country. In the summer of 1992 his mother, accompanied by his younger brother, went to Texas to urge him to return home.[19]

Ismail's role in the conspiracy was essentially a passive one. He represented the only Kuwaiti thread in the flimsy tissue of Ramzi Yousef's identity.

Abdul Basit Revisited

During his flight from Islamabad to New York, Ramzi Yousef reinforced the Abdul Basit identity, informing the FBI agents who accompanied him that "he was born in Kuwait on April 27, 1968, to Pakistani parents."[20] (When he signed his waiver of rights before talking to the FBI, however, he did not do so as Abdul Basit, but used the name "Baluch"—of course, it would have been risky for him to attempt to reproduce Abdul Basit's signature.) He also claimed that he had "remained in Kuwait until the age of twenty, attending and completing his education there at the al-Faheel [sic] School."[21]

The real Abdul Basit did in fact attend the Fahaheel secondary school in Kuwait, the same school as Ismail attended for two years. Fahaheel is a neighborhood just south of Kuwait City inhabited by immigrant workers and their families, including many Palestinians.

Within three weeks of his September 1992 arrival in America, Yousef began calling people outside the New York area who had attended the Fahaheel school. The *Dallas Morning News* identified four such individuals on Yousef's telephone records: two shared a residence in Houston; a third lived in San Marcos, Texas; and a fourth was a Kuwaiti businessman, living in Kuwait. As the man in San Marcos explained, "They mostly reminisced about Kuwait."[22]

But Yousef had come to the United States for the sole purpose of carrying out a major terrorist attack. He was known in his immediate environs as "Rashid the Iraqi." Why would he expand his circle of contacts in the United States beyond the group of conspirators and draw unnecessary attention to his presence? The least likely explanation is that Yousef really was Abdul Basit, nostalgic about his childhood in Kuwait. The most likely explanation is that Yousef's calls were an element in developing a false identity as Abdul Basit.

Yousef first spoke with Eyad Ismail on December 9, 1992. Ismail later told Jordanian authorities that when Yousef first called him, he asked, "Don't you know me? Try to think way back in your memory." Speaking of Kuwait, Yousef said, "Don't you remember a neighbor of yours, with shifty eyes?"[23] Yousef succeeded in convincing Ismail that he was Abdul Basit and that Ismail should know him.

Yousef called Ismail many times subsequently, including a call on February 9, 1993, at 1:00 A.M. Later that day, Ismail bought a ticket back home with a stop in New York. As Ismail told Jordanian authorities, he was planning to go home anyway. On February 21, Ismail left Dallas for New York. Yousef was at the airport to meet him and took him to the apartment at 40 Pamrapo Avenue, where the bomb was being made. The place smelled of chemicals; Ismail found it unpleasant and went to stay at a hotel in Brooklyn, with which he was familiar from his days as a New York cabby.[24]

Yousef told Ismail that he had developed a new business, delivering shampoo. He wanted Ismail to carry out the first delivery. Ismail, having lived in America, would have a better idea how to deal with Americans, Yousef maintained. So around 9:00 A.M. on February 26, 1993, Yousef and another man knocked on his door. Ismail had yet to take his morning shower, so he told them to wait.[25]

The other man was introduced to Ismail as "Shakib." Shakib was very dark, with curly hair, and he spoke broken Egyptian Arabic, according to Ismail.[26] Of course, an Arab raised in an Arabic-speaking country would not speak broken Arabic; he would speak the language properly. It would appear that Shakib probably was not Arab, but of some other ethnicity.

They rode in the van with Ismail driving and Yousef giving directions. Eventually they arrived at an underground parking lot. Yousef told Ismail where to park. There was a red car waiting, with two men in it. Yousef spoke with them in a language Ismail did not understand—neither Arabic nor English, both of which he understood. Yousef lit a fuse that lay on the floor of the van, between the front seats, and the three piled into the car. But a truck blocked the exit. They honked the horn madly before the truck moved, and they succeeded in leaving the garage before the bomb went off.[27]

Ismail stood trial with Yousef in the summer of 1997 for the World Trade Center bombing. The government maintained that Ismail drove the bomb-laden van and that he had done so out of loyalty to his old high-school friend Abdul Basit. Ismail maintained that he drove the van believing that there were boxes of shampoo in it, not boxes of explosives.[28] Would Ismail have had a plausible motive for helping Yousef carry out a major terrorist attack?

The prosecution of Ismail rested on the claim that Yousef was Abdul Basit. Nevertheless, that was not what the government maintained when it issued its first indictment against Ismail in September 1994. That indictment also included the charges against the two other men who were indicted fugitives at the time—Ramzi Yousef and Abdul Rahman Yasin. Among the counts in that indictment was the charge that Yousef fraudulently obtained the Abdul Basit passport. That charge assumed that Yousef was not Abdul Basit; one cannot really obtain one's own passport fraudulently, as is further discussed in chapter 16.[29]

Moreover, Abdul Basit was three years older than Ismail. Even if Yousef were Abdul Basit, he would not have been in the same class with Ismail and the two would have spent only one year—the 1985–1986 academic year—in the Fahaheel school together. After that Abdul Basit was studying in Great Britain, and when he returned in June 1989, Ismail was no longer living in Kuwait. It is hard to imagine that Abdul Basit and Ismail could have had more than a superficial acquaintance.

Finally, Yousef had no need to bring in another person to drive the van. He and the individual introduced to Ismail as "Shakib" could drive the van; they had driven it to Ismail's hotel in Brooklyn. Yousef would thus have brought into the conspiracy, for a wholly superfluous purpose, the only individual in the plot who knew Yousef by what was ostensibly his real identity, as Abdul Basit Mahmood Abdul Karim. Moreover, Yousef fully expected that person to be arrested, as he made clear to the FBI agents who accompanied him to America.

That account, taken at face value, makes no sense at all. However, it makes a good deal of sense if Yousef had assumed the identity of a dead man, Abdul Basit Mahmood Abdul Karim.

Ismail had been brought into the conspiracy precisely in order to be arrested and to confirm that assumed identity under interrogation.

Ismail was convicted in the Trade Center bombing and sentenced to life in prison. As in the case of Ahmad Ajaj, an American jury presented with a medley of Middle Eastern defendants in a case of terrorism proved unable to sort out the individuals' differing degrees of involvement. Perhaps they were simply unwilling to take responsibility for acquitting an individual whom the government had charged with such a heinous crime. They convicted them all.

The existence of other individuals involved with Ramzi Yousef and the other conspirators—including the airplane pilot, the person in "Waly Samar's" dormitory room, the individuals mentioned by Ismail, and perhaps even the microbiologist—raises serious questions about the Clinton administration's decision to handle terrorism primarily as a judicial issue. If the Trade Center bombing had been officially recognized as a foreign intelligence operation, investigators would undoubtedly have looked for more complexity in the operation and worked harder to find higher-ups. But since the bombing was said to be the work of a "loose network," almost everyone involved "on the ground" was equally a perpetrator to be arrested and convicted. As a consequence, a few people are in jail who should not be; those perhaps capable of causing the most serious harm in the future are free; and others who seem to have been involved remain completely unknown.

The government's handling of the Trade Center bombing could not and did not end the terrorism. Rather, the Trade Center was soon followed by another bombing attempt. And the government's handling of that would prove, in its own way, equally problematic.

14

Attempted Bombing of the United Nations Building and Other New York Targets

After the World Trade Center bombing, Emad Salem, the informant whom the FBI had dropped in July 1992, returned, this time prepared to work with the FBI on its own terms. His handlers from eight months earlier met him and heard his complaints that if authorities had listened to him, the Trade Center bombing would never have occurred. Salem also complained about the FBI's attitude toward him: "I'm one of the suspects. . . . I saw my picture on the board, number one suspect." The FBI agent reassured him, "Ah, you were crossed out though."[1]

Salem and the FBI soon made their peace with one another, and another undercover operation ensued—essentially, what Salem had wanted to do the summer before, when he had been unwilling to follow FBI procedures. This time Salem agreed to cooperate with those procedures, and he helped the New York Muslim extremists build a very large bomb—or what they thought was a bomb.

Salem's claim is worth examining; if the FBI had listened to him, he contended, the Trade Center bombing would not have occurred. Egyptian president Husni Mubarak made a similar statement on the eve of a visit to the United States in April 1993—namely, that the Trade Center bombing could have been averted if the United States had heeded Egypt's warnings. But if Salem, with his links to Egyptian intelligence, had not insinuated himself among the New York extremists and suggested that he could help them make bombs, they might never have tried to do so, as they did not know how to. They might well have limited themselves to the activities in which they had been engaged before they met him.

Fourteen people, along with Shaykh Omar Abdul Rahman, were originally indicted in the conspiracy to bomb the United Nations and three other New York targets—the Federal Building and the Lincoln and Holland Tunnels. (See illustration 14–1, the plans of the conspirators.) The centerpiece of the twenty-count indictment was the charge of seditious conspiracy, of which Shaykh Omar was the alleged leader. But it also included charges of a bombing plot (of the Trade Center); murder (of Meir Kahane); and attempted bombing (of the United Nations, the Federal Building, and the two tunnels).

The seditious conspiracy law grew out of a post–Civil War statute used against supporters of the Confederacy who continued to resist the federal government. It was amended in 1918 during the Red Scare that followed the Russian Revolution, and it was used to prosecute socialists and deport immigrants of suspect political beliefs.

The seditious conspiracy law was abandoned thereafter for many decades, until the Reagan administration revived it to prosecute radical leftists, white supremacists, and Puerto Rican nationalists. Such charges pressed against the first two groups failed to convince juries in 1988 and 1989.

The sedition law is unusually broad. It states:

> If two or more persons in any State or Territory, or in any place subject to the jurisdiction of the United States, conspire to overthrow, put down, or to destroy by force the Government of the United States, or to levy war against them, or to oppose by force the authority thereof, or by force to prevent, hinder or delay the execution of any law of the United States, or by force to seize,

ILLUSTRATION 14–1
DIAGRAM OF THE PLANS OF THE ATTEMPTED UN BOMBERS

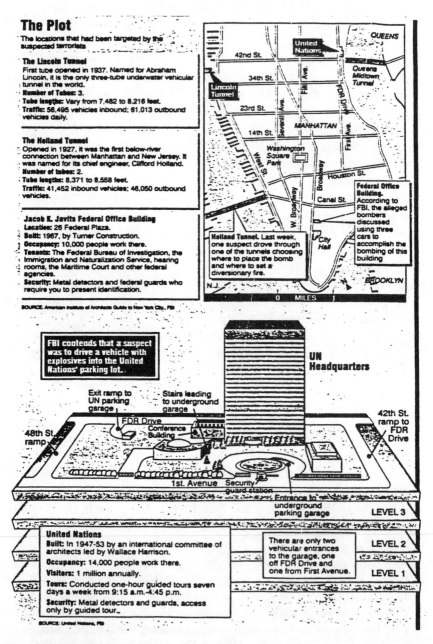

SOURCE: *New York Newsday*, June 25, 1993.

take, or possess any property of the United States contrary to the authority thereof, they shall each be fined not more than $20,000 or imprisoned not more than twenty years, or both.[2]

As the *New York Times* explained, "Because the law does not require the government to prove that the defendants committed any overt acts to further their conspiracy—or even that they knew of all the acts the others committed—some criminal defense experts say the law comes perilously close to punishing people for their beliefs or speech."[3]

When the first arrests in the second New York bombing conspiracy were made early on the morning of June 24, 1993, authorities in Washington could not decide whether to arrest Shaykh Omar or not. Local investigators waited all that day and into the night for a decision from the Justice Department. None came, and Shaykh Omar was not among those arrested.

A barrage of criticism followed, particularly from New York politicians. Shaykh Omar was picked up and jailed a week later, on July 2, on immigration charges. (In March, an immigration judge in New Jersey had ordered him deported—for bigamy. But Shaykh Omar was not imprisoned then.)

U.S. authorities told the Egyptian government that they would deport Shaykh Omar to a third country unless Egypt made an extradition request. Shaykh Omar, for his part, intended to fight to remain in America, in what could have become a protracted legal battle.

Egyptian officials were angry. They considered that Washington was passing them a hot potato. Efforts to extradite Shaykh Omar might prove tricky. The Egyptian government would have had to rely on an 1874 extradition treaty between the United States and the Ottoman Empire, which had sovereignty over Egypt at that time. But the treaty specifically forbade extradition for political crimes. Moreover, if the extradition request had been pressed successfully, Shaykh Omar's detention in Egypt might have provoked more unrest there. Although repeatedly arrested and detained, Shaykh Omar was twice acquitted in Egyptian civil courts—once in 1984, in conjunction with the assassination of Anwar Sadat, and again in September 1990, two months after his arrival in the United States, on charges of inciting violence and attacking police.

Still, Cairo did not want Shaykh Omar sent to a third country, much less to a potentially hostile state, whence he could continue his activities against the Egyptian government. Egyptian officials preferred, above all, that Shaykh Omar remain in an American jail. In mid-August, the Afghan prime minister offered Shaykh Omar refuge in Afghanistan, and the shaykh indicated that he might accept the offer and abandon his efforts to remain in America.

But that did not happen. The U.S. position on Shaykh Omar shifted, and he was indicted for seditious conspiracy. That charge was developed by the Office of the U.S. Attorney for the Southern District of New York—sensitive, presumably, to local political pressures.

Of the fifteen original defendants in the trial of Shaykh Omar et al., two turned state's evidence; a third, Muhammad Abu Halima, the younger brother of the convicted Trade Center bomber, was tried separately on more limited charges. A fourth was released after serving a brief sentence, as his role was minor: Earl Gant, a black convert to Islam, had agreed to supply explosives, but he had believed they were destined for Bosnia and, in any case, never supplied them.

Emad Salem directed the construction of the bombs, but the organizer of the plot—the figure who chose the targets and enlisted others into the conspiracy—was a thirty-two-year-old Sudanese with U.S. residency, Siddig Ibrahim Siddig Ali. Like Nosair, Siddig Ali "found" Islam in America; in Sudan he had enjoyed the discothèques that existed before Hassan Turabi's fundamentalist party seized power there in 1989.

Siddig Ali was among Nosair's more prominent supporters during the Kahane assassination trial, and he often served as a translator for Shaykh Omar. He was also close to Mahmud Abu Halima. Before the Trade Center bombing, Abu Halima confided to Siddig Ali that he was involved in a bombing plot and asked Siddig Ali to assist him in a test detonation of the explosive material. Immediately after the bombing, Abu Halima went to Siddig Ali to boast that his group had been responsible. As Siddig Ali later told Emad Salem, Abu Halima "became so happy" that he "hugged him hard" and then, along with Abu Halima's younger brother, helped him flee the country.[4]

Besides Shaykh Omar, Siddig Ali, Ibrahim El Gabrowny, and El Sayyid Nosair, those indicted in the second bombing conspiracy included three more Egyptians, four more Sudanese, a Palestinian, two American blacks, and a Puerto Rican, the last three being converts to Islam.

There were links between some of those convicted in the Trade Center bombing and some of those convicted in the second New York bombing conspiracy. For example, some of those charged with the attempted bombings participated in paramilitary training with some of those convicted in the Trade Center bombing.

But although the Trade Center bombing was among the counts in the seditious conspiracy charge against Shaykh Omar et al., *none of those convicted in the second bombing conspiracy was charged with actually having participated in the Trade Center bombing*. The government was quite clear about that. It explained, "No defendant in the *Rahman* trial was charged with executing the World Trade Center bombing. The bombing was charged as an overt act of the conspiracy count in the *Rahman* case, and none of the *Rahman* defendants are named in that overt act."[5] Thus, the government's charge was that some of those who participated in the Trade Center bombing, like Muhammad Salameh, were inspired to do so by Shaykh Omar's fiery rhetoric. Hence its inclusion as an overt act in the conspiracy.

But the FBI had not wanted to indict Shaykh Omar. It had wanted to deport him. It did not think the evidence justified pressing criminal charges against him. Moreover, as New York FBI chief Jim Fox later recognized, the trial and conviction of Shaykh Omar fed a fraudulent notion regarding the source of the terrorist threat to America, the so-called loose networks. And the widespread acceptance of that view has left this country vulnerable to the far more serious danger posed by terrorist states.

Fox once described to me the meeting in which the decision to indict Shaykh Omar was made. Attorney General Janet Reno chaired the meeting. Fox was present, as were Mary Jo White, the U.S. attorney for the Southern District of New York; an official from FBI headquarters; and a representative from the Justice Department. Fox argued strongly against indicting Shaykh Omar. The other FBI representative argued mildly against doing so. White argued mildly in favor, and the representative from

Janet Reno's office argued strongly in favor. Five minutes before the end of the hour-long meeting, Reno rapped her knuckles on the table and said, "Okay, we'll indict him." As Fox remarked much later, "I wish I had spoken up."[6]

The decision to include the Trade Center bombing as an element in the conspiracy charges against Shaykh Omar et al. blurred the differences between the two groups, as it blurred the fact that the attempted bombing plot was an FBI undercover operation, while the Trade Center bombing had foreign state sponsorship. In neither case did the New York Muslim extremists make a bomb without very significant outside assistance.

State Involvement with the Extremists

The first plan that occurred to Siddig Ali in the wake of the Trade Center bombing was murder. Even before Mahmud Abu Halima had fled New York for Egypt, Siddig Ali talked to him about assassinating Mubarak during an upcoming visit of the Egyptian president to the United States. Assassinating Mubarak was something that Shaykh Omar wanted very much to see done, having urged that on Emad Salem back in 1991. After Egyptian authorities arrested Abu Halima at his family's home in Egypt on March 16, 1993, Siddig Ali decided it was time to do just that.

But Siddig Ali's scheme was less than well considered. Abdul Muhammad Haggag was a thirty-four-year-old Egyptian who had come to the United States twelve years before. He lived in the same Jersey City apartment building as Shaykh Omar and had participated in the extremists' paramilitary training.

Siddig Ali wanted Haggag to pose as a member of the staff of the Waldorf Astoria, where Mubarak was to stay during his visit to New York. Then, with grenades hidden under his hotel uniform, Haggag was to blow himself up, along with the Egyptian president. Haggag, however, responded by suggesting that he knew of someone else who had already set out to kill Mubarak.[7]

The Origins of the Conspiracy: The Plotting of Siddig Ibrahim Siddig Ali. On April 27, Siddig Ali asked Salem to help with " 'an operation' against an American military target," which he subsequently described as an armory on Manhattan's East

Side.[8] Some ten days later, however, on May 7, Siddig Ali changed his mind. He told Salem that he wanted to attack the United Nations. He also explained that he could obtain diplomatic license plates that would allow them to drive a vehicle into the UN parking garage.

With an eye toward implicating Shaykh Omar, some two weeks later, on May 18, Salem asked Siddig Ali what the shaykh had said about his idea. Siddig Ali claimed that Shaykh Omar had pronounced it an Islamic duty, "very much allowable . . . a must."

But five days after that, when Salem put the question to Shaykh Omar directly, he got a different answer. Shaykh Omar told Salem, "It would not be forbidden, but it would muddy the waters for Muslims." Salem then asked, "Do we do it?" "No, find a plan to inflict damage on the army, the American army, because the United Nations would harm Muslims, harm them tremendously."[9]

But Siddig Ali had set his mind on a bombing campaign. And he was not satisfied with attacking just the United Nations. Once the idea of a bombing plot took hold, he kept adding more targets. On May 18, Siddig Ali explained to Salem that he also wanted to bomb New York's Federal Building. When Salem asked Shaykh Omar whether the Federal Building was a legitimate target, the shaykh replied, "A little bit later, we'll talk about this." Salem explained that the operation had already begun. Shaykh Omar responded, "It doesn't matter. Slow down. Slow down a little bit. The one who killed Kennedy [presumably Sirhan Sirhan] was trained for three years"—a cryptic remark, as it is unclear that Shaykh Omar knew anything about Robert Kennedy's assassination.[10]

Thus, Shaykh Omar's view of Siddig Ali's schemes constituted less than a ringing endorsement, although that is not to say he was a benign figure. Rather, he was oriented toward achieving political goals in Egypt. He was a senior figure in the so-called Islamic Group. He wanted to assassinate Mubarak, and he repeatedly justified the killing of Western tourists in Egypt on the grounds that they carried AIDS and went there for "fornication, drinking intoxicants, gambling, and usury." As Ahmed Sattar, an Egyptian follower of the shaykh, explained, "His concern is with Egypt."[11]

Ten days after Siddig Ali first proposed bombing the United Nations and the Federal Building, he added another target. He told Salem he wanted to carry out three bombings within a ten-minute period. "This will drive the whole world crazy! This will make all America on stand by!" he proclaimed.[12] And two weeks later, by June 14, Siddig Ali had four targets—the Lincoln and Holland Tunnels and the two buildings.[13] As he told Salem, he wanted to complete the project quickly, "to shake the thrones of tyrants here."

Finally, as if all that were not enough, in the latter half of June, Siddig Ali told Emad Salem that he had recruited a pilot to bomb Mubarak's residence in Cairo, along with the U.S. embassy there.

How did Siddig Ali think he could carry off the bombing of four major sites in New York, not to mention the attacks in Egypt? One is tempted to suggest that Ali was a very violent individual, but without the practical, calculating qualities that could have transformed his bloody vision into reality.

Still, a key question remains: Why did Siddig Ali decide to switch from attacking a Manhattan armory—which he considered a U.S. military target, the kind of target Shaykh Omar endorsed—to attacking the United Nations? Many Americans accepted it as a logical target, citing UN positions on Bosnia, Somalia, the Palestinians, and so forth. But that reasoning is extremely vague. Indeed, Saddam offered the same sort of vacuous explanation for invading Kuwait. He was, if you believed him, prepared to leave Kuwait, if only the United States would pressure Israel to solve the Palestinian problem.

Those fuzzy explanations explain little. Of course, Muslim extremists (and many Muslims) have grievances. Their very identities, rooted in strict and deeply held religious beliefs, are at odds with the secular societies that dominate the world today. That has been the case for the past one hundred years.

That dissonance may explain their anger and discontent.[14] But only a tiny minority ever undertake violent criminal action on the basis of their existential anger. The sentiment alone is not enough to explain the deed. Moreover, that sentiment—one that sustains a perpetual sense of discontent—leaves those who are prepared to act upon it subject to manipulation by others who are more calculating in their designs.

It is far from clear that the United Nations constituted a logical target for Muslim extremists. The most authoritative Islamic opinion expressed on the subject was offered by Shaykh Omar, who clearly said not to do it. Despite his professed allegiance to the shaykh, Siddig Ali ignored him and lied to Salem about what Shaykh Omar had said.[15] So why did Siddig Ali fixate on bombing the United Nations?

Sudanese Involvement in the Attempted Bombings. There was state involvement in the second New York bombing conspiracy, as the U.S. government knew. Soon after the June 24 arrests, U.S. authorities intercepted a telephone call from Sudan's UN ambassador to Hassan Turabi in Khartoum. Turabi heads Sudan's fundamentalist party, the National Islamic Front, and at that time was the power behind the military regime that formally ruled the country. The UN ambassador protested to Turabi that two people on his own UN mission staff had been involved in the attempted bombing of the United Nations, and that he had known nothing about it. Turabi responded curtly, "Mind your own business."[16]

Sudanese intelligence was using its UN mission as a base from which to penetrate the local émigré population. Siddig Ali himself had two "friends" at the mission—intelligence agents—who offered to provide him the diplomatic plates to get a vehicle into the UN parking garage. During the trial of Shaykh Omar et al., the prosecution mentioned the two men, Siraj Yousef and Ahmed Yousef Muhammad, but identified them only as employees of the Sudanese Mission to the United Nations—as if they were no different from anyone else involved in the plot.[17]

But not only did they have diplomatic immunity from arrest; they were acting, in fact, in an official capacity, on orders from the Sudanese government.

That became public knowledge three years later. In the spring of 1996, the Clinton administration sought to have the Security Council impose sanctions on Sudan for terrorism. Only then did the administration reveal that two Sudanese diplomats had been involved in the second bombing plot. One of them, Siraj Yousef, had left the United States in 1995, and the other, a second secretary, had remained. The United States ordered him expelled in April 1996.[18] As State Department spokesman Jamie Rubin explained, "We told the Sudanese that we will not permit diplomats to use their immunity as a way of allowing New York City to become a safe haven or a base for terrorism. . . . Therefore we requested the departure of Mr. Muhammad from the United States and gave him forty-eight hours to comply."[19]

Why the Delay? At the time of the June 24, 1993, arrests in the second New York bombing conspiracy, *New York Times* columnist Thomas Friedman wrote, "Administration officials said they had *no conclusive evidence that the Sudanese or Iranian intelligence services were involved* with the Muslims recently arrested in the New York area and accused of plotting to bomb the United Nations and other sites [italics mine]."[20] But they did have such information! Not only had they intercepted the incriminating phone call, but the FBI itself was running the conspiracy. It knew that two Sudanese intelligence agents had offered to provide diplomatic plates to Siddig Ali so that he could bomb the United Nations. And in August 1993, the State Department added Sudan to its list of terrorist states. Indeed, as a former State Department official explained, Sudan's involvement in the second New York bombing conspiracy was the "deciding factor" in putting Sudan on the list then.[21]

Why did the Clinton administration wait three years—from June 1993 until April 1996—before explaining that Sudan had had a role in the second New York bombing conspiracy?

The criminal question of the guilt or innocence of those involved in the New York bombing conspiracies was separated from the national security question of state sponsorship or involvement. Indeed, that was implied by what the prosecutors—Andrew McCarthy, who led the prosecution of Shaykh Omar et al., and Gil Childers, who led the prosecution in the first World Trade Center trial—told me during a January 1995 meeting hosted by the New York district attorney's office: "We prosecute individuals. We don't do state sponsorship." Somehow, that was accepted as a legitimate way to deal with the matter. Of course, one can understand the prosecutors' perspective. Their primary job is to convict the individuals they indict, and they are not responsible for national security affairs. Still, it was the separation of the criminal question from the national security question in the first two major bombing conspiracies to occur on Clinton's watch that gave rise to the notion that a new terrorist threat to America had emerged, and that major acts of terrorism were no longer state-sponsored, but rather the work of individuals and loose networks.

And while the Clinton administration dealt with the *criminal* question in the two bombing conspiracies very publicly, through trials, it dealt with the *national security* question quite differently. It took some actions against the states involved in the terrorism, apparently believing that those measures would prevent future terrorism. But the administration never explained to the public why it took the measures it did. In those decisions, made in the first six months of Clinton's first term, lay the makings of a very serious mistake.

In the first place, the administration misunderstood a basic point. It correctly recognized that Sudan could not have been the primary sponsor of the second plot. What Siddig Ali had come to propose was havoc and mayhem, practically open war. But U.S. relations with Sudan were not that bad at the time; after all, Sudan was not even officially designated a terrorist state. The administration therefore assumed that Sudan was fronting for another party. But then, thinking in terms of Islamic fundamentalism, policymakers focused on Iran.

Iran or Iraq? But Iran had no reason to bomb the United Nations. For Iran, the most important consideration in its relationship with the United Nations at the time was UNSCR 598—which favored Iran. That resolution had been the basis for the August 1988 cease-fire to the Iran-Iraq war. It stated that Iraq was the aggressor in that war, and that Iraq owed Iran tens of billions of dollars in reparations. Thus, Iran had no quarrel with the United Nations.

Who, then, did have a quarrel with the United Nations? Under whose auspices was the Gulf War fought and sanctions imposed and maintained? The obvious candidate for sponsoring terrorist attacks on both the United Nations and United States is Iraq. And Sudan's ties with Iraq are closer than they are with

Iran. Sudan and Iraq are both Arab and Sunni-dominated, while Iran is Persian and Shiʻi.

Moreover, Hassan Turabi himself—the de facto head of Sudan's government at the time—has longstanding links with Baghdad. In 1986, several years before he assumed power, Turabi had helped lure to Sudan the most intellectually formidable of Saddam's clerical opponents—the London-based Mehdi al-Hakim. Knowingly or unknowingly, Turabi effectively set him up for assassination. Al-Hakim was gunned down in Khartoum by Iraqi assassins.

Sudan supported Iraq during the Gulf War, and after the war it was, for a while, the only state allowed to fly planes into Iraq. In October 1992, the UN sanctions committee approved the weekly shipment to Iraq of "frozen meat" from Sudan. In February 1993, that permission was expanded to allow twice-weekly shipments—although no one really knew what cargo the planes carried. Much later, it appeared that Iraq might be producing VX in Khartoum.[22] Some observers suspected that the Sudanese planes might also have carried out from Iraq the equipment and chemical precursors to produce the highly lethal chemical agent.[23]

And Iraq and Sudan cooperate over a broad range of issues. Those include extensive coordination between the Ministries of Interior—even a formal cooperation agreement between them. Such an agreement implies a degree of coordination on intelligence matters.

Khartoum is in fact a major center for Iraqi intelligence. Abd al Samad al-Taʼish, a highly placed Iraqi intelligence agent, was Iraq's ambassador to Sudan until the summer of 1998.[24] Al-Taʻish arrived in Khartoum in July 1991 with thirty-five other intelligence officers to establish a major base for Iraqi intelligence operations there, in the wake of the upheaval wrought by the Gulf War.

Almost certainly, Iraq was the hidden hand behind Sudanese intelligence in its dealings with Siddig Ali. Two days after the arrests in that plot, the Iraqi newspaper *Babil*, owned by Saddam's son Uday, praised the planned attack: "The attempt reflects the anger of those who consider that the United Nations no longer represents humanity." It warned that "the United Nations will become the target of all freedom fighters."[25] No other country praised or justified the attack.

My assessment of the sequence of events in the development of that plot is as follows: Sudanese intelligence learned of Siddig Ali's plan to bomb a Manhattan armory through its existing ties with him. It passed that information on to Iraq, and Iraq changed the conspirators' original plan into a far more lethal operation of unprecedented ambition and destructiveness.

The Administration's Response. When the FBI had assembled all the evidence that it needed against the conspirators—most dramatically, a videotape of them mixing what they believed to be explosive material—it arrested them just before daybreak on Thursday, June 24, 1993. Two days later, on June 26, President Clinton launched a cruise missile strike against Iraqi intelligence headquarters—a punishment, he said, for Saddam's attempt to kill George Bush on his visit to Kuwait in April. The day after the U.S. strike, in an article entitled "The Missiles' Message," Thomas Friedman wrote,

> White House officials said today that their only regret about the missile attack on the Iraqi intelligence headquarters was that there was no CNN crew there to broadcast the event live so it could be watched in the Sudan, Iran, and other countries suspected of involvement in terrorism. . . .
>
> They said that when the White House was planning the strike in Baghdad, it had not only the Iraqi audience in mind but also the intelligence services of countries suspected of sponsoring terrorism, like the bombing of the World Trade Center in New York.

That was the first, and to date the only, suggestion that official Washington suspected state involvement in the World Trade Center bombing—which, after all, was supposed to be the work of a "loose network." Friedman continued,

> By attacking the Iraqi intelligence headquarters with cruise missiles, the Administration was trying to signal that those involved in state-sponsored terrorism would be personally targeted in response.
>
> "We were planning all along to retaliate against Iraq, if and when we had conclusive proof that it was

responsible for the assassination attempt on Bush," said one senior official. "But the situation in New York was very much on our mind and an explicit part of the discussions. We hoped that the response would send a message to those who work in the intelligence business that there is a danger, a danger to them personally, if they get involved in state-sponsored terrorism."[26]

The Clinton administration apparently believed that its attack on Iraqi intelligence headquarters would serve as a response to the New York bombing plots.[27] The strike would deter Saddam from further terrorism, while Sudan and Iran, which were thought to be involved in the second conspiracy, would draw the appropriate conclusions.[28] The point was further reinforced when Sudan was added to the State Department's list of terrorist states.

Apart from the off-the-record comments made in Friedman's *New York Times* article, the Clinton administration never publicly acknowledged that New York FBI and other New York law enforcement agencies suspected that Iraq was behind the Trade Center bombing. Why should the U.S. government choose to cover up its suspicions of Iraqi involvement?

The administration had two distinct frameworks for dealing with the terrorist attacks. On the surface, the attacks were handled as a judicial matter. Perpetrators were held responsible, arrested, and brought to justice. The question of state sponsorship, however, was dealt with quietly.

By dealing with state sponsorship in that way, the administration avoided riling the American public—which, if aware of the suspected involvement of countries like Iraq and Sudan in terrorism on U.S. soil, might have demanded that we do a great deal more.

By the time the trial of Shaykh Omar et al. began, the central figure in organizing the bombing conspiracy, Siddig Ali, had turned state's evidence. Yet in the spring of 1993, when confronted with the charge that he was informing to authorities, Siddig Ali had proclaimed,

> The person who does this must be killed, brother. Whoever does this. If I find somebody working for the FBI

against me, by God, I will kill him. By God I will kill
him by my own hands. By God, I will kill him by these
hands.[29]

For nearly two months after turning state's evidence, Siddig Ali
gave the government near-daily briefings. Then, in August 1994,
he turned back again and stopped cooperating. Finally, in January 1995, just as the trial was to begin, Siddig Ali defected once
again, and finally, to the government. As he told a stunned courtroom, "I wish that those who would listen to my plea today would
understand the reason behind this, that I am doing it because I
am convinced today that I was wrong, and that I ask God to forgive me for all my acts."

Siddig Ali is evidently an erratic and unstable character.
The existence of figures like him and Muhammad Salameh,
prominent among the conspirators in the New York bombing
plots, helped give rise to the notion that the World Trade Center
bombing and the subsequent attempted bombings were the work
of passionate amateurs. By this point, however, it should be clear
that that is not so.

15

Ramzi Yousef's Second Bombing Conspiracy

I n January 1995, Ramzi Yousef shared an apartment on the sixth floor of the Dona Josefa apartments in Manila with Abdul Hakim Murad. He had a plan to bomb twelve U.S. airplanes in what was intended to be forty-eight stunning hours of rage. Yousef was in the last stages of implementing that plan. But late in the evening of January 6, while mixing explosives in the kitchen sink, he started a fire. A neighbor reported the fire to the receptionist downstairs, and she sent a guard up to see what was going on. Yousef and Murad assured him that everything was okay. But dirty white smoke was billowing out the window, and the guard called the police. No one answered the telephone at the police station, so the guard called the fire department, which arrived just before midnight.

Meanwhile, Yousef and Murad had left the building. They waited nearby at a karaoke bar until the firemen departed. Then Yousef sent Murad back to the apartment to retrieve his computer and a number of sensitive documents they had left behind. But while Murad was packing up, the police suddenly appeared. Murad was arrested.

Yousef's Airplane Bombs. And Filipino police seized Yousef's computer. They could not read the files, which were encrypted.

But when the computer was turned over to U.S. authorities, they were able to crack the commercially available encryption program Yousef had used. On Yousef's computer they found plans for a campaign to bomb twelve U.S. commercial aircraft flying on routes within Asia and then over the Pacific to Los Angeles and San Francisco. Jelly-like bombs, meant for the American planes, were also found in the apartment.

And Yousef had found a way to get his bombs onto the airplanes. As he boasted to the FBI agents who accompanied him back to America, he could make bombs that could get past the most sophisticated screening machines. Certainly, his bombs could pass through airport metal detectors. They were made of nitroglycerine and contained no metal. According to the plans detailed on his computer, Yousef and four other conspirators, including Murad, were to board airplanes on the first leg of a given Asian flight, say from Singapore to Hong Kong. The conspirators would assemble the bombs on the planes, hide them, and then disembark at the first destination, leaving the bombs to explode as the aircraft flew over the Pacific.

Indeed, Yousef had done a test run of his bomb on a Filipino airplane. It was the first terrorist bombing of an airplane since 1989. On December 11, 1994, Yousef boarded Philippine Airways Flight 434 as "Arnaldo Forlani," a former Italian foreign minister whose name Yousef had taken from a data base of Italian politicians. Yousef somehow smuggled the nitroglycerine explosive onto the plane, perhaps in a bottle appearing to contain contact lens solution. He assembled the bomb in the bathroom and then slipped it into the life jacket under his seat. As the aircraft stopped to pick up passengers on the Philippine island of Cebu, Yousef disembarked. The bomb exploded when the plane was at cruising altitude on the second leg of the Tokyo-bound flight.

The bomb blew a hole in the plane, and a Japanese passenger—Haruki Ikegami, a twenty-four-year-old engineer sitting in the seat Yousef had occupied—was sucked into the hole. A steward and another man tried in vain to pull him up, but he was mortally wounded, nearly cut in two. Two others were wounded. The bomb also damaged the plane's hydraulic controls. It was only with great difficulty that the pilot managed to steer the aircraft and make an emergency landing on Okinawa, forty-five minutes away.

Yousef's bomb was extremely clever and, in its own way, quite sophisticated. It required a great deal of expertise to devise and construct. Yousef used a Casio watch as the timer. As he explained to the FBI, there is a small, empty section next to the alarm on that watch. Inserting an electronic component into the empty space converts the alarm into a timing device. The alarm can then be connected to a small light bulb in which the glass is broken, exposing the filament. When the alarm goes off, a nine-volt battery, acting as a detonation device, sends power to the light bulb, heating the exposed filament that is packed inside specially treated cotton. The explosive-drenched cotton ignites, in turn setting off the nitroglycerine.

Thus, Yousef's bomb was composed of the most innocent-looking items. Security procedures at airports at that time would not have detected such a bomb. And they would not today.

Yousef and Poison Gas. Yousef also told the FBI that he had contemplated assassinating President Clinton when the president stopped off in the Philippines in November 1994 on his way to a summit of the Asian Pacific Economic Conference. Yousef suggested two ways that could be done. One was to down the plane with a missile, either on take-off or on landing. The other alternative was to stop the presidential motorcade by disabling the lead vehicle and then slathering the motorcade with phosgene, a poison gas first used by the Germans in World War I.

In Yousef's apartment, Filipino authorities found a chemistry textbook, *Hawley's Condensed Chemical Dictionary*. The 1,288-page book was first published in 1930. Yousef had the eleventh edition, published in 1987. It is a technically sophisticated book. A considerable background in chemistry is necessary to understand it. And in the *P* pages of the dictionary, Yousef had taped in an insert, a few paragraphs Xeroxed from another textbook, *Advanced Organic Chemistry*. The paragraphs explained how to produce phosgene.

U.S. authorities found the names of some of Yousef's associates on the computer he had left behind in Manila. Ishtiyaak Parker was a twenty-five-year-old South African Muslim, one of thirteen children from a family of Indian tradesmen. Parker had come to Pakistan in 1991 to study at Islamabad's Islamic University. His modest ambition was to take up a position as a Muslim cleric back home after finishing his studies. His name was on Yousef's computer, and through Parker, U.S. authorities succeeded in nabbing Yousef.[1]

Parker had first met Yousef in Islamabad in May 1994. Yousef appealed to him on the basis of his antipathy to Shi'a Muslims. According to friends and family, Parker was "pious, unworldly . . . without great street smarts"; and a fellow South African who studied with Parker in Islamabad suggested he was "a bit docile."[2] Like Muhammad Salameh and Eyad Ismail, Parker seems to have been none too bright and unusually manipulable. Yousef seems to have sought out and cultivated such people.

Yousef succeeded in fleeing the Philippines after the fire. Using a Pakistani passport in the name of "Adam Qasim," he flew to Singapore. A month later, on February 6, he arrived in Islamabad. In the late afternoon he came to the Su Casa Guest House, located just two blocks from the residence of the Iraqi ambassador.[3] Checking in, Yousef presented a Pakistani identity card in the name of "Ali Muhammad Baluchi" from Pasni, a town in Pakistani Baluchistan located 100 kilometers southeast of Turbat, on the Arabian Sea. Yousef made two telephone calls to Peshawar and a few local calls in Islamabad before Parker arrived, around 9:30 p.m., when the two went out briefly for a cup of coffee. Yousef never went out again until the morning—when he was bound hand and foot, under arrest and, after a day of interrogation by Pakistani authorities, on his way to America.

Murad had used the name "Saeed Ahmed" in the Philippines. When Yousef was arrested, a draft of a letter was found with him. It was written in the name of the "Liberation Army"— the same name as was used to claim credit for the World Trade Center bombing. The letter demanded the immediate release of Saeed Ahmed and threatened "the strongest actions" against "all Filipino interests." It claimed the "ability to make and use chemicals and poisonous gas . . . for use against vital institutions and residential populations and the sources of drinking water and others." It also warned, "These gases and poisons are made from the simplest ingredients which are available through the pharmacies. And we could as well smuggle them from one country to another if needed."[4]

Yousef is the only convicted terrorist who has tried to use chemical weapons to kill Americans in large numbers. Subsequently, he contemplated their use to kill President Clinton, and he threatened their use to secure Murad's release. Yousef's expert knowledge of chemistry allowed him to make chemical agents as well as to build bombs—from the enormous explosive involved in the World Trade Center bombing to the small bombs with which he intended to attack U.S. airplanes.

The U.S. government's assertion in the first World Trade Center trial—that the bombmaking manuals that Ahmad Ajaj brought into the country were linked to Yousef's terrorism—was misleading and potentially quite dangerous. It implied that little knowledge was required for the kind of terrorism Yousef undertook and threatened. And it suggested that such knowledge might well be acquired in relatively primitive places, such as the mujahedin camps in Afghanistan. And if that contention were generally accepted, then after a major terrorist attack—whether it involved conventional explosives or chemical agents— authorities would have no good idea of where to look for the conspirators most responsible. Nor would they have any good idea of which parties were most likely to have been behind the attack.

Moreover, on March 8, a month after Yousef's arrest, two U.S. consular officials in Karachi were killed and a third wounded in a professional ambush of their embassy vehicle. The assassinations, occurring one month to the day after Yousef was sent to the United States, were thought to be revenge killings. Despite an exhaustive manhunt, Pakistani officials failed to make any arrests.

The Lifestyles of Yousef and His Co-Conspirators. As U.S. authorities came to understand, Yousef's second plot began in August 1994. It was then that he came to the Philippines with another conspirator, Khalid Shaykh Muhammad, who remains at large. They stayed in Manila for two months. In October the conspirators gathered in Pakistan, where Yousef instructed them in building his airplane bombs. The conspirators included Abdul Hakim Murad, Khalid Shaykh Muhammad, a third individual who remains unidentified, and Waly Khan Amin Shah.

Waly Khan is an Uzbek. The Uzbeks are a Sunni, Turkic people whose homeland lies on the steppes of Central Asia. The area is perhaps most famous as the region where the fourteenth-century Mongol conqueror, Tamerlane, was born. To the north of Afghanistan, Uzbekistan was formerly a republic of the Soviet Union. Since 1991 it has been an independent state.

Notably, none of those arrested and convicted for Yousef's plane-bombing plot were Arabs. Neither were they Muslim extremists.

Those who knew Yousef in the Philippines explained that "he dressed well in Western-style clothing and that he appeared to enjoy Manila's raucous night life."[5]

Raghida Dirgham, *al-Hayat's* UN correspondent, interviewed Yousef while he was incarcerated in New York. A slim, attractive Lebanese woman, Dirgham got the distinct impression that Yousef was not a religious extremist. During the interview, she tried to put Yousef at ease and gain his cooperation with an occasional flirtatious smile. As she wrote, "He responds to smiles, and smiling makes him shy."[6] In her experience, a radical Muslim would have averted his eyes, rather than return her smile.[7]

While I was reading the transcript of the trial for the plane-bombing plot, a woman approached me who said she had been a court stenographer for both Shaykh Omar's trial and Yousef's trial. She observed that Shaykh Omar's trial had been far more interesting. I asked why. Shaykh Omar's trial, she explained, had been about politics, but this trial was not. Yousef was just "a killer."

And she mentioned an audio file on Yousef's computer, played during his trial, of a conversation between him and a woman named Cindy. Cindy cooed to Yousef, "I love you." Yousef responded, "Shut up, you bitch." Indeed, another audio file was mentioned during the trial. Yousef's lawyer objected to its being played for the jury. He explained that it was only profanity: "As I understand, the one tape contains 'fuck, fuck, fuck, fuck, fuck.' "[8]

That is obscenity, and strict Muslims are not obscene. Islam, like all the major monotheistic religions, enjoins sexual modesty. Muslims are forbidden to have sexual relations with women outside of marriage. And very pious Muslims, whether they are politically radical or not, may go to extremes to maintain an atmosphere of strict propriety. Like many very observant Jews, if they find themselves about to be alone in a room with a woman to whom they are not related, they will ask a third person to come into the room or make a point of leaving the door open. And they will not shake such a woman's hand. Even that contact, as casual as it is in the West, is forbidden.

In Manila, Waly Khan had a Filipina girlfriend, Carol Santiago. Khan had a red motorcycle that he was frequently seen riding with Santiago in tow. At the trial, a Filipino described her as "quite beautiful. . . . When I see her she's always wearing these fitted clothes, and she's kind of sexy."[9] Khan shared an apartment with her. Khan was first arrested by Filipino police on January 12, although he managed to escape two days later and was not apprehended until a year after that. Among the items police found when they searched the couple's Manila apartment were condoms.

For a while, authorities monitored the telephone calls of Khalid Shaykh Muhammad. Among other things, he boasted over the phone of his sexual conquests.[10]

If these people are neither Arabs nor Muslim extremists, who are they? How about individuals recruited by Iraqi intelligence from remote and lawless fringes of the Islamic world, largely unfamiliar to Americans?

On October 31, 1994, Yousef left Karachi for the Philippines. Waly Khan departed at the same time, leaving on an Afghan passport under the name "Waly Khan Amin Shah." He traveled through Singapore and Malaysia and changed passports, arriving in the Philippines on a Norwegian passport under the name "Grabi Hahsen Ibrahim." Khan also had a Saudi passport under the name "Farhan al Dusary."

On November 7, Yousef checked into the Casa Blanca Hotel in Manila as "Mr. Ali," saying he was on holiday. The next day, Khan and his girlfriend rented a room for a month in the Dona Josefa apartments, located near Yousef's hotel. Yousef and Khan had invented a phony organization—the Bermuda Trading Company—for which they made their own identification cards. On November 8 they began to purchase chemicals and equipment, using the name of their phony company. In purchasing the chemicals, Yousef called himself "Dr. Paul Vijay."

They first tested their bomb at a movie theater in Manila on December 1. Khan carried out the test with a very small device, 1½ inches in diameter and ¼ inch deep. Khan constructed the bomb during a movie and then left it under a seat, where it exploded during intermission, injuring several people. The U.S. government suggested that their purpose was to assure themselves they could assemble a bomb, place it under a seat, and cause it to explode.

The next day, December 2, Khan left the Philippines for Malaysia. On December 8, Yousef checked out of his hotel and moved into the Dona Josefa apartments, registering as "Naji Haddad" and claiming to be Moroccan. Three days later, he bombed the Philippine Airways plane.

On December 26, Murad arrived in Manila, checking into the Las Palmas Hotel. He stayed there for two days before moving in with Yousef. When he arrived at Yousef's apartment there was, according to Murad, a third man there. He was introduced to Murad as "Abdul Majid," and he moved out that day.

Murad had brought with him several bottles of L'Oréal hair coloring. He used the dye to make his already dark hair darker. On January 4, 1995, Waly Khan returned to Manila. But the fire on the night of January 6 interrupted their plans.

Still, all this—the multiplicity of passports, the frequent changing of names and addresses, and even the hair dye—is illustrative. These people operated at a far more complex and sophisticated level than did the Muslim extremists convicted in the two New York bombing conspiracies.

On his flight back to the United States, Yousef told the FBI agents accompanying him that if it were not for the fire, the bombings would have occurred "within a two-week period."[11] That would have approximated the anniversary of the start of the Gulf War—January 17. Was the plane-bombing plot scheduled to occur on, or near, the anniversary of the war's start perhaps to serve as a bookend, as it were, to the World Trade Center bombing, which occurred on the anniversary of the war's end?

In January 1995 Saddam was in a triumphal mood, as we discussed in chapter 10, anticipating the end of sanctions and asking who would fire the fortieth missile on Israel. Is it possible that Yousef's second plot could have had some connection with sanctions, holding the prospect of hastening their demise?

U.S. authorities found on Yousef's computer a draft of a letter warning of a major attack on "American targets." Apparently, it was to be sent out shortly before the airplane bombings. It stated,

> We the fifth division of the Liberation Army, under the command of Staff Lieutenant General Abu Bakr al-Makki, declare our responsibility for striking at some American targets in the near future in retaliation for the financial, political, and military support extended by the American government to the Jewish state, which occupies the land of Palestine. . . . We will consider all American nationals as part of our legitimate targets because they are responsible for the behavior of their government.[12]

Had the plane bombings succeeded, the warning from the Liberation Army would have linked them with the World Trade Center bombing, as that "army" and "Staff Lieutenant General Abu Bakr al-Makki" had claimed credit for that terrorist strike as well. Since the United States had attributed the Trade Center attack to shadowy Muslim extremists, it would have seemed that the same "group" was behind the second attack. And if the plane-bombing plot had, in fact, destroyed a significant number of U.S. passenger aircraft, Americans would have become tragically aware of the elusive Liberation Army, apparently yet another terrible threat associated with the Middle East. That ominous and mysterious menace had first tried to topple New York's tallest tower onto its twin, and then two years later would presumably have downed a significant number of U.S. airplanes. Could there be any more dramatic threat to the security and well-being of Americans?

Most people have the capacity to worry about only so much at one time. In the face of such a new, terrible threat from Muslim extremists, who would have focused serious energies on dealing with Iraq? At the time, Iraq's friends on the Security Council were strenuously urging that sanctions should be lifted. Might not the sudden re-emergence of an imminent menace in the Middle East—which did not come from Iraq—help to push that process along?

Whether or not such a diversion of public attention was intended, the perception of a dangerous new source of terrorism against America, represented by loose networks of militant Muslim extremists, arguably diminishes the sense of threat that Americans feel from Iraq. The general view is that Saddam has been unable to act against the United States, and the cudgel of anti-Americanism has been taken up by a new, more youthful cohort of terrorists, energized by the passions of militant Islam.

Yet is that so? The notion that there is a new, network-sponsored terrorist threat to America emerged out of three major bombing conspiracies: the two plots in New York and Yousef's plane-bombing plot. None of them, however, can accurately be described as the work of loose networks. Looking at all the evidence, it appears that Saddam was involved in each.

The false and fraudulent theory of loose networks of terrorists began to emerge in the years after the 1991 Gulf War. That theory has greatly benefited Saddam. It allowed him to sponsor acts of terror with virtual impunity. At the same time, it has had the paradoxical effect of diminishing the threat that Americans feel from Iraq, because the threat has been perceived to come from elsewhere.

16

Official Misinformation

In early 1996, a Justice Department official, reading off the FBI arrest card, told me the height of Ramzi Yousef. That was crucial information to confirm that the account given by the Immigration and Naturalization Service of Yousef's height—6 feet—was correct. With the arrest card in hand, he told me that Yousef was "a titch" under 6 feet. He joked about not being able to read the small type, clearly not wanting the information to appear as an official leak. But to pin the matter down a bit more precisely, I asked if, by "a titch," he meant that Yousef was closer to 6 feet or to 5 feet 11 inches. "Six feet," he replied. Yet Abdul Basit Karim was only 5 feet 8 inches.[1]

In February 1996, I traveled to Wales and visited the West Glamorgan Institute of Higher Education, later renamed the Swansea Institute, to meet with two faculty members who had taught Abdul Basit. Brad White, an experienced investigative journalist with whom I was working, arranged that meeting. British authorities had told the people at Swansea not to talk to outsiders about Abdul Basit. But it had been a year since Yousef's arrest, so what was the harm?

The two men with whom I met had a clear memory of Abdul Basit. They described him as a very quiet person. He was quiet with his teachers. And he was quiet among his peers. Not quite the person one would expect to run up $18,000 in unpaid telephone bills, as Yousef had. And not quite the person one would expect to play the leading role, organizing others, in two major bombing conspiracies.

Abdul Basit's teachers also said that they had observed no strong political or religious views in the student. He went to the mosque on Fridays, but there was nothing exceptional about his practice of Islam. He was not an extremist.

Abdul Basit had been a diligent pupil and a hard worker. The students in Swansea's two-year Higher National Degree (HND) program were required to complete a major project in their second year. Abdul Basit had chosen a project that involved computer programming. His teachers still remembered him hunched over his computer, sitting at his work station for long stretches of time. Also, he had wanted to get a master's degree. That was very unusual for students completing the HND. But as one of his teachers remarked, he might have made it.

Abdul Basit's teachers were much surprised at the claim that he was the terrorist who had blown up the World Trade Center and then attempted to bomb a dozen U.S. airplanes. Before Yousef's arrest, they had been shown pictures of him. The pictures were of poor quality, so it was hard to tell much. Still, their opinion was that while there may have been some similarities between Abdul Basit and the pictures they were shown, they did not seem to be the same person.

Also, Abdul Basit's teachers noted that if he had been planning a terrorist career while at Swansea, it would have been easy for him to use his second-year project as a means to further his expertise. He might have done something with a digital watch that could relate to the construction of a timing device, for example. They had seen projects like that. But Abdul Basit's project had no relevance to terrorism.

For his project, Abdul Basit took the school symbol and redesigned it as Islamic art. He then developed a computer program that would draw the picture. As his professors explained, given the state of computer technology at the time, that required a great deal of work. It entailed writing the instructions in computer language for each line the computer was to draw. And that was how his teachers most clearly remembered Abdul Basit— crouched over his computer, laboring on his second-year project.

In fact, the project was so exemplary that Abdul Basit's supervisor kept it. He used it in subsequent years as a model to show students what their second-year projects might be like, and to inspire them. It was later handed over to British authorities. Latent fingerprints are impressions left on absorbent surfaces, in the course of the ordinary handling of material. In the supervisor's opinion, the project should have had Abdul Basit's latent fingerprints on it, and it probably did.

And what about the height of Abdul Basit? His two teachers suggested that he was about 5 feet 6 inches or 5 feet 7 inches tall. "What about 5 feet 8 inches?" I asked. That, after all, was the height on Abdul Basit's 1988 passport, and also the height reported by Tony Geraghty. The two men agreed that Abdul Basit might have been 5 feet 8 inches. I then asked, how about 6 feet? But they said no. The teachers asked each other, "Did you look up at him or did you look down at him?" They agreed that they had looked down at Abdul Basit, and that he was not 6 feet tall. They were confident of that.

Ramzi Yousef would shortly be on trial in Manhattan. I asked the teachers whether, if I flew them to New York and they saw Yousef in the courtroom, they could pronounce definitively whether he was or was not Abdul Basit.

They declined. They said the only way they could be 100 percent sure was to meet with Yousef and talk with him. But, of course, the U.S. government had custody of Yousef. And only the U.S. government could arrange that.[2]

Indeed, the issue of Ramzi Yousef's identity—whether he was Abdul Basit or not—was taken up subsequently by a British journalist, Simon Reeve. Reeve claimed that Yousef's real identity was Abdul Basit. Reeve had excellent access to U.S. officials and he, too, gave Yousef's height as 6 feet. But he misrepresented Abdul Basit's height, claiming that Abdul Basit's 1984 passport, issued when he was sixteen years old, indicated that he was 5 feet 8 inches tall and that "it is not unheard of for late developers to sprout a few inches in their late teens."[3] But Abdul Basit's 1984 passport stated that he was 4 feet 7 inches at the age of sixteen. It is Abdul Basit's 1988 passport—issued when he was twenty—that indicates he was 5 feet 8 inches tall (see appendixes A and B). And Abdul Basit could not have grown four inches between the ages of twenty and twenty-four—as he would have to have done, if Ramzi Yousef were really the individual born "Abdul Basit Mahmood Abdul Karim" in Kuwait on April 27, 1968. [4]

Thus, the identity of the most formidable terrorist in U.S. history remains a mystery. The simple step of having people who knew Abdul Basit from the period before Iraq's invasion of Kuwait meet with Yousef in prison to confirm the key question of whether he is or is not Abdul Basit has not been taken. Nevertheless, five months after Yousef's arrest, a senior administration official in Washington asserted that "the bulk of the evidence to date" argued that Yousef was Abdul Basit, even as he conceded that "few are 100 percent certain."

Moreover, if Yousef really were Abdul Basit, wouldn't it have been important for the government to have used that name during Yousef's prosecution? That way, all those who knew Abdul Basit would understand that he was a terrorist. They could come forward and help explain what happened in his life between June 1989 and September 1992 that turned him into a terrorist.

I asked a New York prosecutor why, if the government believed that Yousef is Abdul Basit, did it not try him as Abdul Basit? He replied, "It doesn't matter what we call him. We just try a body."[5] A former Senate Intelligence Committee staffer later reacted to that remark by commenting, "That's bizarre."[6]

If Abdul Basit were indeed Ramzi Yousef, that would mean that in June 1989 he was a pleasant, diligent, hard-working young man whose most immediate professional ambition was to acquire a master's degree. But by September 1992, three years later, when Yousef arrived in America, he had become a veritable monster. His ambition to get a master's degree had been cast aside in favor of a project to kill 250,000 innocent human beings by toppling New York's tallest tower onto its twin, even using cyanide gas as a lethal enhancement. And he had picked up an extraordinary knowledge of chemistry, a subject not taught at the Swansea Institute. Moreover, the fact that Abdul Basit's file was tampered with during Iraq's occupation of Kuwait must be sheer coincidence, as must be the other incongruities discussed earlier, such as Yousef's never once signing his name as Abdul Basit.

Is that likely? I don't think so. And the way that some U.S.

government officials dealt with this information, as it emerged, suggests that they didn't think so, either.

Ramzi Yousef was arrested and returned to the United States on February 7, 1995. By that time, many people had been told of the significance of his identity—whether or not he was, in fact, Abdul Basit. In November 1994 I had discussed this with Jim Fox, who agreed that the question of Yousef's identity was indeed the "smoking gun" and had told New York FBI about it.[7] Subsequently, I conveyed the information not only to the New York prosecutors but also to officials in Washington, including an individual at the White House. The Clinton administration's response to this issue may have been predictable, but it came as a shock to me. It claimed that Ramzi Yousef was Abdul Basit.

The last indictment issued by the U.S. attorney's office in New York prior to Yousef's arrest came in September 1994. The charges still included the following count: "Ramzi Ahmed Yousef obtained a Pakistani passport in the name 'Abdul Basit' that was procured by means of a false claim and statement, and otherwise procured by fraud and unlawfully obtained."[8] That, of course, is tantamount to saying that Yousef is *not* Abdul Basit; one can scarcely obtain one's own passport fraudulently. But in the next indictment, issued shortly after Yousef's arrest, that charge was dropped. The government ceased to claim that Yousef had fraudulently obtained the Abdul Basit passport.

Moreover, after U.S. officials began to claim that Yousef was Abdul Basit, they asserted that the identification was based on fingerprints from Abdul Basit's file in Great Britain, not Kuwait. That was reported by *U.S. News* on February 20, 1995: "There [in Great Britain] he had been fingerprinted. After the trade center bombing, sources say, FBI agents matched prints belonging to Yousef to those of Basit."

That could not have been so. Tony Geraghty had already learned that Abdul Basit had no criminal record in Great Britain, and that meant that the British did not have his fingerprints. In the United States, when a foreigner obtains a "green card" his fingerprints are taken; in Great Britain, alien registration does not involve taking fingerprints. A person's fingerprints are retained in a British file only if he is convicted of a crime.

Not every U.S. official claimed that Yousef was Abdul Basit at that time. Nearly two months after Yousef's arrest, the well-

connected *New York Times* reported that Yousef's identity remained unknown. In fact, it was one of the "central mysteries" in
the case.[9]

But shortly after Yousef's arrest, I happened to hear that an
official in the U.S. attorney's office in New York claimed that the
fingerprints identifying Yousef as Abdul Basit came from Great
Britain. Perhaps, I thought, there was some mistake. Perhaps
the man had been misinformed; so I called him.[10] He repeated
the claim. I said it could not be, and he should check the files
himself, copies of which were in the possession of New York law
enforcement. Still, he insisted that the fingerprints identifying
Yousef as Abdul Basit came from Great Britain, and he even
asked what more I needed to know, before I would accept what
he was saying. I explained British procedures for registering
aliens. If he maintained that Abdul Basit's fingerprints came
from Great Britain, I wanted to know what crime Abdul Basit
had been convicted of, because I would go to Great Britain and
check the court records myself.

He had no answer. The same man later told me that he had
been referring to latent fingerprints, taken from material Abdul
Basit had handled in Great Britain.[11] But that was not true either. If the British had found latent prints of Abdul Basit match-
ing Yousef's prints, it would have been clear fairly early in the
investigation that Yousef was indeed Abdul Basit—and U.S. authorities would *not* have charged Yousef with falsely obtaining
the Abdul Basit passport, as they did in all four indictments issued from August 1993 through September 1994.[12]

Why did U.S. authorities go to such lengths? The fundamental problem was the way the White House had dealt with the
bombing, compounded by developments related to the trials. At
the time of Yousef's arrest, the trial of Shaykh Omar had just
begun. One cannot overstate the importance of that trial in explaining why the Office of the U.S. Attorney in New York would
suppress information suggesting state sponsorship of the World
Trade Center bombing.

Indeed, after Yousef's capture, as the arrest of the master
terrorist seized the headlines, many journalists called Jim Fox
to get insights from the man who had first supervised the FBI
investigation in New York. Fox pointed them toward Iraqi

involvement in the bombing. But it was not long before he got a call from Mary Jo White, who headed the Office of the U.S. Attorney in New York. White was in charge of the prosecutions in the bombing conspiracies, and she asked Fox to stop speaking with the media. Fox remarked to me, "Boy, does she not want state sponsorship addressed."[13]

As the *New York Times* explained, Mary Jo White is a highly regarded, aggressive lawyer, and "she pressed successfully for the arrest and indictment of [Shaykh Omar] when other law-enforcement figures were reluctant to charge him unless he could be tied to a particular illegal act."[14]

That was the problem. The indictment of Shaykh Omar was dicey—and to be certain of getting Shaykh Omar, the information suggesting that Iraq was behind the Trade Center bombing had to be suppressed. Indeed, Shaykh Omar's defense team initially wanted to argue that the Trade Center bombing was an Iraqi intelligence operation.[15] Eventually, the judge (a man known in New York for his ambition to be promoted to a higher court) ruled that whether Iraq was or was not behind the Trade Center bombing was irrelevant to the guilt or innocence of the defendants. But that issue had not been settled at the time of Yousef's arrest. And the prosecution had to consider what the jury might have thought if, during Shaykh Omar's trial, it had emerged in the press that Iraq was behind the Trade Center bombing—and that Sudanese intelligence, and behind that, Iraq had been involved in selecting the targets in the second bombing conspiracy. Would the jury still have convicted Shaykh Omar?

From the outset, the judicial inquiry into the World Trade Center bombing was constrained by a policy decision, made by the White House, to separate the question of the involvement of states from the question of the guilt or innocence of individuals. It was further—and fatally—distorted by strategic and tactical calculations in the service of a different, and competing, prosecution.

17

The Islamic Change Movement

"**O**f *course* that was Iraq. That was a professional bomb. It was not made by a bunch of Saudis sitting in a tent in the middle of the desert." So a senior Saudi official told me in February 1996. He was referring to the bombing of the U.S. training mission for the Saudi National Guard in Riyadh, just three months before.

Chapter 10 discussed at length UNSCOM's April 10, 1995, report, which nailed Saddam for an undeclared biological weapons program. That report raised the serious prospect that sanctions would never be lifted as envisaged in UN Resolution 687—through a clean bill of health from UNSCOM. So, it seems, Saddam took the first steps to prepare for a campaign of force and violence aimed at undermining the coalition arrayed against him.

On April 11, an explicit threat to U.S. forces in Saudi Arabia was published in *al-Quds al-Arabi,* a London-based, Palestinian-owned newspaper that is sympathetic to Baghdad. It was the first such threat to U.S. troops since the Gulf War. As *al-Quds al-Arabi* reported, "An extremist group in Saudi Arabia has threatened to carry out military operations against the 'crusader forces' in the Arabian peninsula, especially U.S. and British forces and what it described as the influential members of the al-Sa'ud family."

The statement had been faxed to the newspaper by the "Islamic Change Movement," and it marked the first appearance of the group's name on the world stage. The statement called on foreign forces to leave Saudi Arabia by June 28, after which they would become a "legitimate target."[1] The next day, Iraq Radio broadcast the threat as it had been published in *al-Quds al-Arabi*.

In early July, *al-Quds al-Arabi* reported that it had received another statement from the Islamic Change Movement. The new statement reaffirmed the June 28 deadline and asserted, "The movement has been making preparations since the first statement. However, it does not mean that the operations will be carried out immediately after the deadline." And it reaffirmed, "The movement will use all available means to move 'the crusader forces' off the peninsula of Islam."[2]

The Riyadh Bombing. The threats were also faxed to the U.S. embassy in Saudi Arabia. And, indeed, on November 13, a 250-pound bomb exploded in Riyadh, killing five Americans and two Indians. The bombing occurred at 11:40 A.M., prayer time for Saudis, lunchtime for Americans. A witness saw two men get out of the van that carried the bomb shortly before the explosion. He was able to describe one of them in sufficient detail to allow authorities to produce a sketch of the suspect. Notably, the man had a mustache but no beard; deeply religious Muslims do not shave, just as deeply religious Jews do not.

It was the first such bombing in Saudi Arabia. And it was the most lethal assault on a U.S. facility in the Middle East since 1983, when a Syrian/Iranian-backed bombing campaign drove U.S. forces out of Lebanon. It was also the worst terrorist attack against the United States overseas since Libya's downing of Pan Am Flight 103 in 1988.

The Islamic Change Movement claimed credit for the bomb. So did two other previously unknown groups. One was the "Tigers of the Gulf," which claimed responsibility in a phone call to Agence France-Presse. The claim of another group, the "Combatant Partisans of God Organization," was first broadcast on Iraqi television.[3]

Indeed, the Iraqi media hailed the attack. *Al-Iraq* ran the banner headline, "Gulf Tigers Shake the Saudi Throne and Deprive Washington of Sleep."[4] Another paper warned that "the huge explosion" was "a very serious message from the Saudi people to the United States and the world."[5] As a Saudi paper observed, "Nobody has paid tribute to this crime, except for the press in Iraq."[6]

Kuwaiti officials immediately suggested that Iraq was behind the bomb.[7] So, too, did the veteran Israeli journalist Uri Dan. Dan reported that the chief of Saudi intelligence told the U.S. ambassador, "The explosive device was far too sophisticated to be assembled locally. We believe it was brought into the country."[8] That, too, is what Hussein Kamil, in exile in Jordan, believed. Citing a source close to Kamil, Dan reported,

"[Saddam is] determined to beat the U.S. and 'topple the White House into the dust,'" Kamil said. "That's one of his favorite expressions. You can judge how sly he is: He wants everybody to think it's the Iranians behind the attack. He is using them as a smokescreen, hoping to fool the world. He will go on trying to seek revenge against his enemies. Top of his list are Saudi Arabia and the U.S."[9]

Indeed, the U.S. ambassador in Saudi Arabia described the bomb as "very sophisticated," reflecting what the Saudis had told him. And he hinted that it was state-sponsored, noting that "Saudi Arabia lives in a bad neighborhood," implying Iran or Iraq.[10]

That was a line of speculation that the Clinton administration was not interested in pursuing. The ambassador's remarks were repudiated the next day by an unnamed embassy source, who said, "At this stage, there is no hard information to point a finger."[11] Indeed, Clinton's first response had been to suggest that the bomb was the work of individuals. On the morning of the attack, the White House issued a presidential statement condemning it as an "outrage" and vowing to "work closely with [the Saudis] in identifying those responsible for this cowardly act and bringing them to justice."[12] Only individuals are brought to justice; states must be punished in other ways. At a press conference later that day, Clinton said, "We have begun the process of determining what happened and who, if anyone, was responsible, if it was not an accident. And we will devote an enormous effort to that."[13] Strangely, Clinton seemed to suggest that something other than a bomb might have sheared off the face of the building.

Although several different Middle Eastern sources suspected Iraq, the notion was alien to much of the American elite. The *Wall Street Journal* considered the possibility that Saddam was responsible, but dismissed it: "Experts doubt he has the capability, after five years under international sanctions, to perpetrate such a well-planned bombing inside Saudi Arabia."[14] Similarly, the *Washington Times* asserted that Baghdad was "hardly likely to engage in anything so provocative while trying to convince the U.N. Security Council that it has at last come clean on its prohibited weapons programs in order to end the U.N. sanctions."[15] But that is American thinking. It is not how Saddam thinks. As Tariq Aziz explained, "People do what you want *when you hurt them*" (emphasis added).[16]

There is no proof that Iraq was behind the Riyadh bombing. Yet Iraq should have been considered a prime candidate, and it was not. The Arab-Israeli peace process was the primary preoccupation of the Clinton administration in the Middle East. It was also the primary preoccupation of Israel's Labor Party government. With several agreements in hand—that between Israel and the Palestinians in 1993, that between Israel and Jordan in 1994, and an expectation that an agreement with Syria was imminent—the climate of euphoria was incompatible with the notion that the war with Iraq was not yet over.

Moreover, a dangerous confusion was creeping into the U.S. understanding of terrorism. A bombing that five years earlier would have been considered state-sponsored no longer was recognized as such. By now, terror was regularly being attributed to ghostly figures: Muslim militants, who had somehow acquired the skills and means once regularly attributed to states.

That, apparently, was the conclusion that U.S. authorities reached about the Riyadh bombing. *New York Times* columnist Thomas Friedman, with excellent contacts in the Clinton administration, wrote a month later that the bomb "was probably an in-house job. That is, the attack was masterminded by Saudis against Saudis. *There is no evidence yet of foreign direction*"[17] (italics added). That, of course, was to discount the bomb's sophistication, and to ignore the threats from the Islamic Change Movement that began the day after the critical UNSCOM report, as well as the celebration of the bombing in Iraq's official media.

In our February 1996 meeting, I also explained to the Saudi official with whom I discussed the Riyadh bombing that Iraq had been responsible for the Trade Center bombing. I showed him the copies of the passports Ramzi Yousef had presented to the Pakistani consulate to obtain the passport on which he fled, pointing out the signatures in the 1984 and 1988 passports—how completely different they were—and telling how Abdul Basit's file in Kuwait had been tampered with to create a false identity for Yousef. The man was stunned. He asked, in disbelief, how the Pakistanis could even have thought to give a passport to someone on the basis of such documents.

I also explained how the Clinton administration had separated the question of state sponsorship, with which it dealt indirectly, from the criminal question of the guilt or innocence of the individuals involved, with which it dealt in trials.

We discussed the danger posed by Iraq's unconventional weapons, as they had been revealed after Hussein Kamil's defection. Was Saddam holding on to that material because he intended to use it? The Saudi responded, "After all we've been through, don't tell me there's anything that that man wouldn't do." And alluding to how Saddam had carried off the invasion of Kuwait, as the whole world watched, the Saudi official cautioned, "Saddam is capable of strategic deception."

In April, the Saudi government announced the arrest of four Saudis, all Sunni extremists, for the Riyadh bombing. An aide to the Saudi official with whom I met assured me that even with the arrests, the Saudi government was *not* claiming that the Riyadh bombing was *not* state-sponsored. In May, shortly before the four men were executed, their "confessions" were broadcast on Saudi television. All four "wore the broad beards that are an emblem of devout Islamic faith," but to Western diplomats in Riyadh, the confessions seemed "scripted" and "surreal."[18]

Were those individuals really involved in the bombing? The executions created friction between U.S. and Saudi authorities.

The Americans had wanted to question the suspects before they were executed, but the Saudis had refused to allow it. If the four men had really been involved in the bombing, it is hard to understand why the Saudis would have refused the U.S. request to interview them.

Why did the Saudis not accuse Iraq if they believed that Iraq was behind the bomb? Perhaps because if the United States was not prepared to back up the charge, there was no point in making it. The Saudis would only anger Saddam without being able to punish him for the bombing.

The al-Khobar Bombing. The failure to recognize that the Riyadh bombing was a serious act carried out by a serious enemy led to a much bigger attack seven months later.[19]

From June 21 through June 23, 1996, the Arab League held its first summit since the Gulf War. The meeting of the Arab leaders in Cairo represented a heightened degree of coordination among Saudi Arabia, Egypt, and Syria following the election of the Likud Party leader Benjamin Netanyahu as Israeli prime minister. The Saudis were prepared to support Syria in a tough line against Israel in exchange for Syrian support for a tough stance against Iraq. And Egypt was prepared to go along with both.

Iraq was the only Arab country not invited to the summit, which adopted a very hard line against Baghdad. The June 23 summit communiqué called for maintaining sanctions on Iraq. It also demanded that the regime comply with "all the relevant Security Council resolutions," and it affirmed that the regime was "solely responsible for the suffering of the Iraqi people."

The Iraqi media immediately lashed out at the United States, Saudi Arabia, and Kuwait. *Al-Thawra* charged that the summit communiqué "was written with Arabic letters but with American sentences and phrases paid for by the Saudi and Kuwaiti dirty money."[20] *Al-Qadissiya* warned, "Before it is too late, the Arabs should rectify the sin they committed against Iraq when a number of Arab armies joined the thirty-state coalition and participated in the aggression against it."[21] *Al-Iraq* wrote, "Both agent regimes in Saudi Arabia and Kuwait, the enemies of the Arab nation, instigated the Americans, Zionists, and smaller agents to strike at Iraq and destroy it."[22]

And on the night of June 25, just two days after the summit ended, a bomb twenty times larger than the Riyadh bomb exploded outside the base that housed the pilots enforcing the no-fly zone over southern Iraq. Nineteen U.S. servicemen died in the al-Khobar bombing.

A new, unknown group, the "Legion of the Martyr Abdullah al-Huzafi," immediately claimed credit, as did the Islamic Change Movement two weeks later. Yet the kind of statements issued after the Arab summit by Iraq's government-controlled media are what is called "a threat environment" in the intelligence business. It is considered a meaningful indicator in the investigation of terrorism.

Uri Dan, citing Israeli counterintelligence sources, wrote, "They point a finger directly at Saddam Hussein, who is still seeking revenge."[23] Subsequently Dan repeated the claim, quoting a U.S. intelligence official: "The finger of suspicion points straight at Saddam."[24]

A prominent defector, General Wafiq Samarrai, who headed Iraqi military intelligence during the Gulf War, also suggested that Iraq was behind the bombing. As early as February 1995, shortly after his defection, Samarrai explained that following the invasion of Kuwait, Saddam established a secret committee to carry out terrorism, including terrorism in Saudi Arabia. Samarrai was a member of that committee. But, as Samarrai explained, "the plans came to nothing" because of the speed and intensity of the war.[25] Following the al-Khobar bombing, Samarrai noted that the attack resembled plans drawn up by that committee to explode bombs near buildings where American soldiers lived. He also noted that the committee had never been disbanded.[26] Later, Richard Perle, assistant deputy secretary of defense under President Ronald Reagan, also suggested that Iraq was the likeliest suspect in that bombing.[27]

Indeed, the Iraqi media responded to the al-Khobar bombing as it had to the Riyadh bombing. *Al-Thawrah* said that the bomb "will fully convince Washington that keeping its units in the Arabian Peninsula will be like sitting on a mine or volcano."[28] "Mother of Battles Radio" warned, "The U.S. administration has a chance to leave. If it thinks it can cope with such operations, then it should send more coffins to Saudi Arabia."[29] A Kuwaiti paper noted "the smug satisfaction—indeed, scarcely concealed glee—with which the Iraqi media reported the bombing and the fact that the American soldiers who were killed were helping to enforce a 'no-fly zone' over southern Iraq."[30]

President Clinton responded to the second attack as he had to the first. Shortly after the bombing, the White House issued a presidential statement vowing, "The cowards who committed this murderous act must not go unpunished. Within a few hours, an FBI team will be on its way to Saudi Arabia to assist in the investigation. . . . We are ready to work with [the Saudis] to make sure those responsible are brought to justice."[31] Two days later, at a G-7 summit in France, Clinton repeated himself: "We will not rest in our efforts to discover who is responsible, to track them down and to bring them to justice."[32]

Again, the very tenor of the president's remarks suggested that the bombing was the work of individuals, rather than a state. There was some speculation that the al-Khobar bombing was linked to the Riyadh bombing. Indeed, the day after the bombing a Pentagon spokesman asserted, "The starting assumption is that it was a domestic group."[33] But which group? That was never revealed.

Subsequent speculation also encompassed possible state sponsors. Syria was suggested.[34] But that idea was quickly nixed by the Saudi ambassador in Washington.[35] Saudi-Syrian relations were good, and the Saudis needed Syria's help in dealing with Iraq in Arab affairs. Iran was also suggested prominently as a candidate.[36]

Iraq, however, was scarcely mentioned by official Washington. Of course, the last thing the Clinton administration would have wanted—with the presidential election looming in November—was for Americans to think that Saddam was behind that bomb. But the failure of many others even to suggest Iraq as a suspect reflected the extent to which the Gulf War seems to have been forgotten.[37]

By the fall, the U.S. investigation had come to focus on Usama bin Ladin, the renegade Saudi millionaire and Sunni extremist who, in the 1980s, fought the Soviets in Afghanistan. In 1991, bin Ladin shifted his operations to Sudan. He remained in Khartoum until the summer of 1996, when he left for Afghanistan, after the Sudanese government, under U.S. and Saudi pressure, asked him to leave.

According to a report by Bill Gertz, the highly regarded defense correspondent for the *Washington Times*, bin Ladin had arranged for the explosive C-4 to be shipped to the Persian Gulf shaykhdom of Qatar, whence it was smuggled into Saudi Arabia. "Intelligence reports from 1995 said Mr. bin Ladin met with an Iraqi intelligence officer in Sudan, along with an Egyptian extremist and a Palestinian who was an expert on car bombs, [U.S.] officials said." Officials also said that "bin Ladin was in contact with Iraqi intelligence agents while based near Khartoum" and that he had "contacted Iranian intelligence officers in Afghanistan about seeking political asylum," even though such contacts were "unusual" because of the Sunni-Shi'a divide. Still, "the CIA [had] no clear evidence that either Iraq or Iran was involved in the Saudi bombings, the officials said."[38]

Thus once again, in the fall of 1996, the administration seemed poised to blame Saudi dissidents for another major bombing attack, as it had the year before. Of course, that was very problematic for the Saudis. It raised questions about the country's internal stability and made the opposition to the government seem far more capable than otherwise it was—if the bombings were not the work of Saudi dissidents, or dissidents acting on their own.[39]

That was the fundamental political divide between the United States and Saudi Arabia. The Clinton administration had an interest in an outcome that did not involve finding a state sponsor, lest it be obliged to take action against that state. And Iraq was the worst possible candidate.

At that point, as the United States began to suggest that bin Ladin was behind the al-Khobar bombing, the Saudis publicly leaked information from their own investigation, even before they had presented it to U.S. authorities.[40] Some 200 Saudi Shi'i had been arrested following the bombing, which occurred in the eastern province, a Shi'i-inhabited region. Of that number, 40 were kept in detention. From their interrogation, Saudi security forces came up with the names of a number of people from the Saudi Hizbollah, said to be involved in the bombing. The Saudis held in custody the man who had driven the oil tanker that carried the 5,000-pound bomb, or so they claimed. Another man,

Ahmed Mughasil, said to be the political mastermind, was in Tehran.⁴¹ A third figure, Jaafar Shulkhat, had fled to Syria, where Syrian authorities said he had died in prison.⁴² And a fourth, Hani al-Sayigh, was in Canada. Needless to say, the Saudi interrogation had involved torture, and the information was of questionable reliability.

But the Saudi leak of its own investigation successfully halted U.S. speculation about Usama bin Ladin. The Saudi information suggested that Iran had been behind the bombing, with the possible connivance of Syria. However, when the Saudis made available to U.S. authorities what they were willing to share, the Americans pronounced themselves dissatisfied. FBI director Louis Freeh complained, "I don't have enough information to make strong findings or conclusions." What he had been given he derided as "hearsay information in many cases."⁴³

And the Saudi information never led to any clear conclusions. That was true regarding both the question of state sponsorship and the particular individuals whom the Saudis had identified as participants in the bombing. Indeed, when the reliability of that information was put to the test, it resulted in a very embarrassing situation for U.S. law enforcement.

The man whom the Saudis had identified as a participant in the bombing who had fled to Canada, Hani al-Sayigh, sought political asylum there. Canada rejected al-Sayigh's bid for asylum, and then sought to deport him. It could have deported him to Saudi Arabia, but al-Sayigh did not want to return there. To avoid being deported to his native country, he struck a deal with the FBI and was sent to the United States in June 1997. Al-Sayigh maintained that he had had nothing to do with the al-Khobar bombing, but he claimed that he had been involved in another plot that was never carried out. Al-Sayigh said that in 1995, an Iranian intelligence officer had ordered him to carry out surveillance against possible U.S. targets in Saudi Arabia, and that in December 1995 he had traveled to a southwestern region of that country to determine the availability of weapons and explosives for use against U.S. targets.⁴⁴ He agreed to plead guilty in exchange for a ten-year sentence.

But al-Sayigh's claim was improbable on its face. If he had been involved with Iranian intelligence in planning a serious attack against a U.S. target, the Iranian government would have been quite unlikely to allow him to go to a country where he could easily be picked up by U.S. authorities. Indeed, once he was in the United States, al-Sayigh denied that he had been involved in such a plot. U.S. authorities had no evidence of the plot, beyond his own statements. So they had to drop those charges. Nor were they any better positioned to prosecute al-Sayigh for involvement in the al-Khobar bombing. They had no reliable information to present in a U.S. court, and the Saudis were unable to provide them such information. Once it was clear that no criminal charges would be pressed against him, al-Sayigh asked for political asylum in the United States. That was denied, and he was deported to Saudi Arabia.

It seems that the Saudi information was thoroughly unreliable, as Louis Freeh had complained. The individuals whom the Saudis arrested and named as conspirators in the al-Khobar bombing may well have been involved in something like clandestine political-opposition activity, with the support of Iran. But if they had really been involved in the bombing, then between the FBI and the Saudi security forces, it probably should have been possible for them to develop sufficiently credible information to present in a U.S. court to serve as the basis for a trial. After all, to know which individuals were involved in a terrorist attack is to know a great deal. Authorities can then conduct a very narrow and focused investigation. That neither al-Sayigh nor any of the other individuals the Saudis have accused has been brought to trial—whether in the United States or in Saudi Arabia—raises serious doubts about the credibility of the Saudi claims.[45]

The Damascus Eight, or the D-8, as it came to be known, was an Arab grouping that arose out of the Gulf War. It included the six states of the Gulf Cooperation Council—Saudi Arabia, Kuwait, Bahrain, Qatar, the United Arab Emirates, and Oman— along with Egypt and Syria, the most important Arab states that stood with Saudi Arabia against Iraq in the 1991 war. Following the war, the idea arose that Egypt and Syria might continue to help provide protection against Baghdad, but neither Saudi Arabia nor Kuwait was particularly interested. They did not want large numbers of Egyptian and Syrian troops stationed on their territory. After all, Egypt and Syria might develop designs on their oil, just as Saddam had. They were much more comfortable with U.S. protection, which would also be more effective. Thus, despite the D-8's formal existence, it was not a meaningful organization in the first years after the Gulf War.

But following Hussein Kamil's defection, as it became understood just how dangerous Saddam remained, the gulf states took a renewed interest in the D-8. The organization began to reconvene on a regular basis.

On December 28 and 29, 1996, the D-8 foreign ministers met in Cairo. Following their session, they issued a statement that "reiterated full compliance with the resolutions of the Arab Summit Conference in Cairo on June 21–23, 1996." And "expressing complete sympathy with the Iraqi people, and putting full blame on the Iraqi regime for their plight, the ministers called on Iraq to implement the UN resolutions concerning its aggression against Kuwait and to take the necessary steps to free all detainees, return all property, and cooperate with the compensation process and with the international committee assigned to eliminate weapons of mass destruction."[46]

Two days later, on December 31, a bomb exploded in a bus in Damascus, killing nine people and injuring forty-four. It was the first terrorist bombing in the Syrian capital since the early 1980s, when Damascus ruthlessly crushed an Islamic campaign backed by Iraq and Jordan. Following the New Year's Eve blast, Damascus immediately blamed Israel for the attack. Syrian vice president Abdul Halim Khaddam asserted, "It is a political explosion, not just a group of people committing an ordinary crime. . . . We are convinced that Israel is behind it. . . . And the day will come when we make public all the facts."[47]

But the Islamic Change Movement claimed credit for the bombing. And as Uri Dan, citing British and Israeli intelligence sources, wrote, "The sly fox Assad is well aware that Iraq's dictator (whom he calls 'the snake Saddam Hussein') is almost certainly the mastermind."[48]

That was the last bomb for which the Islamic Change Movement ever claimed credit. In fact, it was the last that was ever heard from the "group." Was it in fact a group at all?

The most likely interpretation is that the Islamic Change Movement was a name invented by Iraqi intelligence to threaten or claim credit for bombings. Political developments in 1995— UNSCOM's April 10 report nailing Iraq for an undeclared biological weapons program, and Hussein Kamil's August 8 defection—ensured that sanctions would not be lifted, as envisaged in UNSCR 687. Saddam was determined to retain both his entire biological program and the other unconventional programs that were revealed after Kamil defected. It would have been clear to him that the Security Council might never vote to lift sanctions. So he apparently turned to a strategy of intensified violence and intimidation, in an attempt to erode the coalition against him.

Nothing can be proven about the authorship of the bombings in Saudi Arabia. Compared with the terrorist attacks that were followed by arrests and trials in U.S. courts, very little information is available about the attacks for which the Islamic Change Movement claimed credit, particularly the last. Nevertheless, the Saudi official whom I interviewed after the first bomb was unequivocal in his claim that Iraq was responsible. The second attack, significantly, targeted the very pilots who flew over southern Iraq. It was also the view of Uri Dan that Iraq was behind all three bombs.[49] And they all occurred, or were threatened, in close proximity to events related to constraining Iraq. The threats by the Islamic Change Movement served as a way of hinting at responsibility for the bombings without being so clear as to make retaliation likely. And because the Saudis picked up on the hints and on other indications of Iraqi sponsorship, while the Americans did not, strains emerged between the two countries.

18

Other Terrorist Conspiracies

C hapter 12 reviewed the series of crises that Saddam drove, starting in October 1997, that culminated in the termination of UNSCOM's presence in Iraq. The second of those crises ended on February 23, 1998, when Secretary General Kofi Annan reached an agreement with Tariq Aziz on the inspection of the so-called presidential sites. And even as that crisis ended, Saddam had the next one in mind.

The narrative below is derived from two quite different sources. The first is the U.S. government indictment of Usama bin Ladin and fourteen others for the August 7, 1998, bombings of the U.S. embassies in Nairobi, Kenya, and Dar Es Salaam, Tanzania. The second, in italics, is a chronology of developments in the confrontation with Baghdad from the resolution of the second crisis to the start of the third, when Baghdad suspended UNSCOM inspections on August 5, 1998.

The Prelude to the Embassy Bombings

On February 22, 1998, Khalid al Fawwaz made many calls to the office of *al-Quds al-Arabi*.[1] Fawwaz is a Saudi extremist who handled media relations for bin Ladin in London. Fawwaz is among those indicted for the U.S. embassy bombings.

The next day, February 23, *al-Quds al-Arabi* published a

statement from the World Islamic Front.[2] The statement was signed by bin Ladin and Ayman al-Zawahiri, head of a radical Egyptian Islamic group. Both were subsequently indicted for the embassy bombings. In addition, the leader of another radical Egyptian organization signed the statement, as did the heads of a Pakistani and a Bangladeshi movement. It was the first appearance on the international scene of the World Islamic Front. It was unclear what caused the five figures to come together to issue the statement, which focused almost exclusively on U.S. actions against Iraq.

The statement read, "The Arabian Peninsula has never—since God made it flat, created its desert and encircled it with seas—been stormed by any forces like the crusader armies spreading in it like locusts, eating its riches and wiping out its plantations." It cited "three facts that are known to everyone," the third of which involved U.S. support for "the Jews' petty state." But the first two focused on Iraq:

> First, for over seven years the United States has been occupying the lands of Islam in the holiest of places, the Arabian peninsula, plundering its riches, dictating to its rulers, humiliating its people. . . . The best proof of this is the Americans' continuing aggression against the Iraqi people using the Peninsula as a staging post. . . .
>
> Second, despite the great devastation inflicted on the Iraqi people by the crusader-Zionist alliance, and despite the huge number of those killed, which has exceeded 1 million . . . despite all this, the Americans are once again trying to repeat the horrific massacres, as though they are not content with the protracted blockade imposed after the ferocious war or the fragmentation and devastation. So here they come to annihilate what is left of this people and to humiliate their Muslim neighbors. . . .

And the statement called on Muslims to kill Americans:

> The ruling to kill the Americans and their allies—civilians and military—is an individual duty for every Muslim who can do it. . . . We—with God's help—call on every Muslim who believes in God . . . to kill the Americans and plunder their money wherever and whenever they find it. We also call on Muslim ulema, leaders,

youths, and soldiers to launch the raid on Satan's U.S. troops and the devil's supporters allying with them.[3]

The Confrontation over UNSCOM

On April 17, UNSCOM issued its semiannual report on Iraq's weapons. UNSCOM said that it had made no progress over the past six months, punctuated as they had been by two confrontations with Baghdad. On April 27, the Security Council reviewed sanctions on Iraq and voted to keep them in place.

On May 1, Iraq's Revolutionary Command Council and Iraq's Ba'th Party leadership, known as the Regional Command, issued a joint statement in the form of a letter to the Security Council. Bellicose and angry, the letter demanded the lifting of sanctions. It was all but a declaration of war. It began, "Wars in general happen primarily because of a maximum feeling of injustice." It asserted, "Iraq has been subjected to injustice of which the annals of history speak of nothing similar to it." And it quoted the Koran: "To those against whom war is made, permission is given to fight, because they are wronged, and truly Allah is most powerful for their aid." It concluded, warning, "This prompts us to competently adhere, after relying on God, to our rights, sound path, and to great jihad for the sake of the rights of our people, our nation, and mankind. . . . The inability of the Iraqi people to see a lifting of sanctions after eight years . . . will lead to dire consequences."[4]

The indictment of bin Ladin includes a section entitled "Preparation for the Bombings of United States Embassies." It begins, "On or about May 4, 1998, the defendant Khalfan Khamis Muhammad applied for a Tanzanian passport in the name "Zahran Nassor Maulid."[5] Muhammad participated in the bombing of the embassy in Tanzania.

On May 7, one of bin Ladin's military commanders, Muhammad Atef, who was also indicted in the embassy bombings, sent Fawwaz, in London, a statement issued by the "Ulema Union of Afghanistan." It denounced the U.S. Army as the "enemies of Islam" and called for "jihad against the United States and its followers." *Al-Quds al-Arabi* published the statement a week later.[6]

On May 17, the Iraqi cabinet reiterated the leadership's

May 1 letter. Referring to a Security Council statement about Iraq's weapons programs, the cabinet said, "This statement does not meet the minimum level of Iraq's rights following the immense sacrifices it has made. Iraq is still waiting in accordance with the concepts included in the open letter it addressed to the UN Security Council on 1 May 1998."[7]

Muhammad al-'Owhali rode on the truck carrying the bomb in Nairobi. On May 17, using a passport in the alias "Khaled Rashed," he traveled from Yemen to Karachi, arriving there May 18.[8] From Karachi, he went on to Afghanistan to join bin Ladin.

On May 29, bin Ladin issued a statement entitled "The Nuclear Bomb of Islam," under the banner of the World Islamic Front. In it he said, "It is the duty of the Muslims to prepare as much force as possible to terrorize the enemies of God."[9]

The individual who drove the bomb-laden truck to the U.S. embassy in Nairobi is known as "Azzam." On June 19, Azzam traveled from Karachi to Nairobi on a false passport under the name "Gihad Ali."[10] Also in June, the conspirators purchased a truck in Tanzania for use in the bombing operation, and they rented a house there. They also purchased the truck that Azzam drove for the Kenya bombing.

On June 23, the RCC and the Ba'th Party leadership repeated their May 1 statement. On July 17, Saddam gave his annual speech marking the anniversary of the coup that brought the Ba'th to power. Saddam warned,

> *The enemies of Iraq will be mistaken if they imagine that they can deceive the people who are fortified by all factors of national life and . . . who were scorched by the fire of their enemies. . . . We assert that the letter addressed to the Security Council and the UN Secretary General by the joint meeting of the Party command and the Revolutionary Command Council on 1 May 1998 is not only a cry of protest, but it represents will power and an alternative strategy. The command and the council will meet at a later date to conduct an in-depth discussion of this issue and deal with it. . . . After that only what is right shall prevail.*[11]

On July 21, the Iraqi leadership again reiterated its May 1 letter and warned of "dire consequences" if sanctions were not lifted. Nine days later, the RCC and Ba'th leadership met and issued a statement in advance of a critical visit by UNSCOM chairman Richard Butler. The statement reviewed Iraq's "cooperation" with UNSCOM, the list of weapons UNSCOM had destroyed, and the sites in Iraq that UNSCOM monitored. And the statement asked, "Why should Iraq put up with all this, if the siege continues and has no forseeable end?"[12]

Meanwhile, in Africa, the conspirators continued their preparations. On August 2, Muhammad al-'Owhali arrived in Nairobi from Pakistan. He used the same false passport to enter Kenya that he had used ten weeks earlier to travel from Yemen to Pakistan.

Muhammad Sadeek Odeh, a Palestinian, had moved to Kenya in 1994, where he established himself in Mombasa, on Kenya's Indian Ocean coast, and worked as a fisherman. In early August 1998 Odeh, along with other members of bin Ladin's organization, al-Qaeda, traveled from Mombasa to Nairobi.[13] From August 2 through 6, Odeh stayed at the Hilltop Hotel in Nairobi, together with other members of al-Qaeda.[14]

On August 3, Richard Butler arrived in Baghdad. From the start of his first meeting with Tariq Aziz, it was clear that something was up. Iraq had five video cameras running, to record the exchanges. Aziz asked Butler for his assessment of the situation as it had developed since he last visited Baghdad six weeks earlier. Butler replied that UNSCOM had made no progress because Iraq had given it virtually none of the information or material it had asked for. Aziz replied that he disagreed, but he would give Butler "the definitive answer of the leadership of the government of Iraq" in a second meeting later that evening.[15] *Subsequently, Aziz told Butler that Iraq had complied with UNSCR 687. And "your only duty now . . . is to leave this room and go back to New York and tell the Security Council that Iraq is disarmed."*[16] *Butler and his delegation left Iraq the next day.*

The following day, on August 5, Baghdad announced "Suspension Day": that is, the suspension of UNSCOM inspections. Iraq's National Assembly met in an extraordinary six-hour session. It recommended that the government "stop dealing with UNSCOM and put an end to its work in Iraq."[17] *Later that day, the RCC and the Ba'th Party command met and issued a joint statement.*

The Iraqi leadership's August 5 statement began with the same Koranic verse that ended its May 1 letter: "To those against whom

*war is made, permission is given to fight, because they are wronged,
and truly, Allah is most powerful for their aid." And it stated,*

> Iraq has suffered throughout eight difficult years from
> an unjust and unfair blockade that is unprecedented in
> modern history. As the whole world knows, this blockade
> led to the death of millions of Iraqi citizens. . . .
>   Iraq has agreed on the UN Security Council resolu-
> tions and implemented them throughout seven years of
> hard and strenuous work . . . which was explained by the
> statement issued on 30 July 1998. . . .
>   In addition to previous warnings over the past years,
> we have been frankly, clearly, and truly warning over
> the past three months—since 1 May 1998—against the
> continuation of this situation and stressed that the lead-
> ership and people of Iraq cannot stand the continuation
> of this situation. . . . These serious and true warnings
> were neglected.[18]

*That was the last time any official Iraqi source ever referred to
the May 1 letter. The statement concluded by announcing that
Iraq was suspending weapons inspections.*

The next evening, August 6, Muhammad Odeh and another
man charged in the embassy bombings, Fahd Muhammad Ali
Msalem, boarded a Pakistani International Airways flight from
Nairobi to Karachi.[19] A third conspirator, Ahmed Khalfan Gai-
lani, also left Nairobi that night for Karachi on a Kenya Airways
flight.[20]

In the early hours of August 7, before the embassy bombings,
"claims of responsibility were sent by facsimile to London, England, in the name of the 'Islamic Army for the Liberation of
the Holy Places' for further distribution by co-conspirators."[21] At
10:30 A.M., local time, on August 7, a truck bomb exploded at the
U.S. embassy in Kenya, killing 213 people and injuring more
than 4,500. Ten minutes later a second truck bomb exploded,
outside the U.S. embassy in Tanzania, killing 11 people and injuring at least 85.

Later that day, President Clinton denounced the "abhor-
rent" and "inhuman" terrorist bombings. And he pledged, "We
will use all the means at our disposal to bring those responsible
to justice, no matter what or how long it takes."[22]

Once again, Clinton responded to a major terrorist attack—this time the simultaneous bombing of two U.S. embassies—as if it were the work of individuals, rather than a state. Iraq figured prominently in the initial media speculation about the bombings because of the conflict over UNSCOM, but the administration soon put out word that bin Ladin was behind the attacks. And it seems he was. But that still leaves a key question: Did bin Ladin act alone, or was he acting in cooperation with a terrorist state, specifically Iraq?

The administration scarcely addressed that issue. From London, Ambrose Evans-Pritchard of the *Daily Telegraph* characterized the U.S. response to the bombings as, "If in doubt, blame Muhammad." And he suggested, "Rather than dwelling on the fevered *fatwas* of bin Ladin, who is reportedly holed up in the Hindu Kush, it might be worth dwelling on the specific threats issued by the government of Iraq, an entity that still has a sophisticated intelligence service and that has not given up its quest for weapons of mass destruction."[23]

Helle Bering of the *Washington Times* wrote similarly. She recalled the old view of terrorism that prevailed before the Trade Center bombing—that it was largely state-sponsored—noting that during the Cold War, "terrorism was a favored tool of the Soviet Union, which trained and funded networks all over the globe, from Germany's Red Army faction to the Irish Republican Army to the Italian Red Brigades and the PLO." And she remarked that the embassy bombings "happened at the same time that Iraq announced its decision to cease cooperation with the UN weapons inspectors. . . . Coincidence? Perhaps, but not likely."[24]

Indeed, why couldn't the bombings have been the work of both bin Ladin and Iraq? Bin Ladin, a Saudi, was furious when, after Iraq invaded Kuwait, the Saudi government turned to the United States, invited U.S. forces into the country, and even financed the war against Iraq. He viewed it as "treason."[25] Moreover, after the war, U.S. forces did not go home. They stayed in Saudi Arabia and continued the war against Iraq, although on a much-reduced scale. It was over that continued U.S. presence that bin Ladin broke with the Saudi government and began to support its London-based opposition. Given the extensive Iraqi intelligence presence in Khartoum, the report (discussed in chapter 17) claiming that there was contact between Iraqi intelligence and bin Ladin while he lived there—although sketchy—is probably true. Indeed, one later contact between bin Ladin and Iraqi intelligence was quite open. In December 1998, a senior Iraqi intelligence official, Faruq Hijazi, visited him in Afghanistan, and that was widely reported.[26]

Although the Clinton administration first suggested it would deal with the embassy bombings as it had dealt with the previous terrorist attacks—by bringing the perpetrators to justice—that soon changed. The administration claimed it had intelligence suggesting that more strikes were planned, and the World Islamic Front did indeed begin to threaten more attacks. So on August 20, the United States launched sixty-six cruise missiles at bin Ladin's training camps in Afghanistan and thirteen missiles at a site in Khartoum involved in the production of the chemical agent VX—or so senior U.S. officials said.

Following the strikes, Clinton addressed the nation. He affirmed, "Our target was terror. Our mission was clear: to strike at the network of radical groups affiliated with and funded by Usama bin Ladin, perhaps the preeminent organizer and financier of international terrorism in the world today."[27]

Later that day, National Security Council adviser Sandy Berger spoke about the U.S. strikes. He explained that bin Ladin was behind the embassy bombings, but he added, "We have not ruled out that others share responsibility, and we are looking into every possibility."[28] Subsequently, it was reported that the White House had considered "targets in three countries: Afghanistan, Sudan, and one nation that U.S. officials declined to identify."[29] Five months later, as Operation Desert Fox began, the United States took the precaution of temporarily closing thirty-eight embassies in Africa.[30] Had the White House really failed to recognize that the embassy bombings occurred just two days after Baghdad angrily suspended UNSCOM inspections, and that the threats issued by bin Ladin had paralleled those issued by Baghdad?

The El Shifa Pharmaceutical Plant. When, on the day of the U.S. retaliatory strikes, Sandy Berger explained the strike in Khartoum, he asserted, "We know with great certainty [that it] produces essentially the penultimate chemical to manufacture VX nerve gas." The plant, Berger said, was run by a Sudanese enterprise "called the Military Industrial Complex." Bin Ladin was a "financial contributor" to the enterprise, and was therefore linked to the VX facility.[31]

But it soon emerged that the facility in Khartoum was the El Shifa pharmaceutical plant. The administration came under strong pressure to justify why it had targeted that site. It responded by providing information linking the plant to Iraq and to suspected Iraqi production of VX.

As the *New York Times* explained, "The United States believed that senior Iraqi scientists were helping to produce elements of the nerve agent VX at a factory in Khartoum that American cruise missiles destroyed last week."[32] The key evidence linking the Khartoum plant to VX production was a soil sample taken in December 1997 from the ground outside the facility by the CIA's Counterterrorism Center.[33] It contained the rare chemical "EMPTA," a VX precursor. And Iraq, according to a senior U.S. intelligence official, was the only country known to have used EMPTA for that purpose.[34]

Indeed, from Khartoum, the *New York Times* reported that back in 1991, soon after the Gulf War, Iraq had struck a deal with Sudan. Baghdad would provide financial help and expert assistance to Sudan, in exchange for which Sudan would allow Iraqi technicians to use its facilities. "Struggling Sudan got economic help, and Iraq was able to move ahead with chemical weapons far away from the gaze of UN weapons inspectors."[35]

UNSCOM was indeed concerned that Iraq had transferred some chemical weapons research and production capacity to Sudan. But UNSCOM did not think that the plant the United States struck was involved in such activities. Rather, its suspicions focused on "a smaller, more heavily guarded facility elsewhere in Khartoum."[36] U.S. officials were familiar with that site and had considered striking it. But unlike the El Shifa plant, it was located in a residential neighborhood and they were concerned about civilian casualties.[37]

Yet if Iraq was suspected of producing VX in Sudan, why had the administration said and done nothing about that before? The UN Security Council had authorized the El Shifa plant to sell medicine to Iraq. Possibly, Sudanese or Iraqi intelligence used the plant at some point to repackage VX salts, a stable form of the agent, so that the material could be made to look like pharmaceutical products and shipped to Baghdad. That, perhaps, could have explained the presence of a VX precursor in the soil sample taken from the area around the facility.

In any event, it turned out that a Sudanese-born Saudi businessman, Saleh Idris, owned the El Shifa plant. He had purchased it in March 1998. The administration claimed that he, too, was part of bin Ladin's network. The *Washington Post* quoted one U.S. official as saying, "What we're learning about [Idris] leads us to suspect that he's involved in money laundering, that he's involved in representing a lot of bin Ladin's interests in Sudan."[38] Similarly, the *New York Times* reported, "Officials say U.S. intelligence has received reports that Idris launders money for international Islamic groups, and that he also has a stake in a company in Sudan that is 40 percent owned by the Military Industrial Corporation, a government entity that the United States says controls Sudanese chemical weapons development."[39] Following the embassy bombings, President Clinton signed an executive order that froze any U.S. assets owned by bin Ladin and his associates. Under the terms of the order, the United States seized $24 million that Idris held in an account at the Bank of America in Great Britain.

But Idris was not a terrorist. And he hired a powerful Washington law firm, Akin, Gump, Strauss, Hauer & Feld, to defend him. A well-known private investigative agency, Kroll Associates, was tasked with looking into Idris's background. And the head of Boston University's Department of Chemistry was assigned to test samples from the bombed factory and the grounds around it.

Exhaustive chemical testing revealed no traces of EMPTA in or around the factory. Moreover, Kroll Associates found no evidence linking Idris to bin Ladin or to any other terrorist, "and lots of evidence to the contrary."[40] And at that point, early in 1999, Idris sought to reach an understanding with the U.S. government. But the government would not meet with his lawyers or hear the reports.

So Idris sued the U.S. government. He presented his evidence in court. And when the government's response was due, Idris's lawyers got a call from the Justice Department. The

United States had unfrozen Idris's assets, rendering any legal action moot. As the editors of the *Washington Post* concluded, "The U.S. government shot from the hip in accusing Saleh Idris of a link to terrorism, and missed badly."[41]

Other Audiences? Clearly, the United States was the primary target in the embassy bombings, but there may well have been a message meant for others as well. The Arab members of the anti-Iraq coalition would have closely followed the threats that Baghdad issued in the confrontation over UNSCOM. They would have noted the bellicose and angry statements Baghdad made on August 5, as it announced the suspension of weapons inspections. So what would they have thought when, less than forty-eight hours later, two U.S. embassies were bombed simultaneously? And what would they have thought when U.S. officials quickly attributed the bombings to bin Ladin and his loose network of Muslim extremists?

The United States is very strong relative to Iraq. The Arab members of the anti-Iraq coalition are weak relative to Iraq. They are not in a position to punish Saddam. If he can act against America with impunity, what might he do to them if he chose?

The embassy bombings suggest that bin Ladin was working with Iraqi intelligence. That discovery would have been somewhat shocking to the Saudis, if they had not recognized it before. And aspects of the attack seemed intended to intimidate them specifically. A fictitious organization, the "Army for the Liberation of Islamic Holy Places," claimed credit for the bombings. Of course, the two most important Islamic holy places are located in Saudi Arabia. Thus, the ostensible aim of the "organization" was to liberate sites on Saudi soil. Moreover, when the "army" claimed responsibility for the attacks, it asserted that two Saudi nationals had carried out the bombing in Kenya.[42]

The administration's handling of the two terrorist bombings of U.S. facilities in Saudi Arabia strained ties between Washington and Riyadh, as was discussed in chapter 17. Quite possibly, the way the attack on the two U.S. embassies was executed may have been intended to produce a similar effect. The administration's predictable response—to blame individuals rather than states—would have served as a pointed reminder to the Saudis of the hesitancy of their senior partner in the continuing confrontation with Baghdad.

19

"International Radical Terrorism" and Other Theories of Terrorism

The Clinton administration maintains that a new terrorist phenomenon has emerged, starting with the World Trade Center bombing. That certainly would seem to be so. But why does the new terrorism exist? The administration claims that there is a new form of Islamic jihad, carried out by individuals rather than states. Steven Simon and Daniel Benjamin, former senior director and director for counterterrorism in the National Security Council, write about "the new face of terrorism." They cite the Trade Center bombing as an example of this new terror. They explain that "highly motivated" individuals were behind this phenomenon, moved "by a world view in which they are the vanguard of a divinely ordained battle." An earlier generation of terrorists sought to achieve political goals and thus kept their violence "carefully calibrated," Simon and Benjamin assert. But the new terrorists "are not constrained by secular political concerns. Their objective is not to influence but to kill, and in large numbers—hence their declared interest in acquiring chemical and even nuclear weapons. It is just this combination—

religious motivation and a desire to inflict catastrophic damage—
that is new to terrorism."[1]

Similarly, Secretary of Defense William Cohen has written
of the "grave New World of terrorism," in which "traditional no-
tions of deterrence and counter-response no longer apply."
Cohen, too, cites the Trade Center bombing as an example of the
new terror. And he warns that the new terrorists are capable of
carrying out a biological attack of such devastating magnitude
that U.S. authorities would be hard-pressed to deal with the consequences. "Hospitals would become warehouses for the dead
and dying."[2]

But are we sure? Are we sure that a new kind of terrorism
exists? If the administration is wrong in its claim that there is a
new terrorist threat, represented by individuals rather than
states, then it could be giving Saddam Hussein, or any other
state sponsor of terror, a license to kill on an enormous scale.

Indeed, the administration's view of terrorism is now widespread and commonly held, including among law enforcement
officials. After any major terrorist attack, even an unconventional one, these are the authorities who would be responsible for
determining who was behind it. If they believe that almost any
group of Muslim extremists has the motive and the ability to
carry out a major attack, how can they prevent such an attack
before it occurs, and where will they look afterward to find those
responsible? Moreover, how will authorities know if there are not
some individuals among the "group" they end up pursuing who
are, in fact, foreign intelligence agents, who provided the direction and expertise for others? Thus, if a terrorist state should
want to kill Americans in large numbers and to have a good
chance of getting away with it, that state should make its attack
look like the work of Islamic zealots. It could even leave behind
a few violent, but witless, dupes to be arrested and stand trial—
just as happened in the Trade Center bombing.

International Radical Terrorism

The FBI provided one of the more authoritative, official explana-
tions of what is supposed to be the new terrorism. The Terrorism
Research and Analytical Center (TRAC) of the National Security
Division of the FBI prepares the bureau's annual reports on ter-

rorism. They are known as the TRAC reports. In the 1994 TRAC report, the FBI elaborated on the nature of the new terrorist threat. The report characterizes the new terrorism as "International Radical Terrorism," or "IRT." It portrays the Trade Center bombing as the paradigm of the new terrorism and seeks to explain, in particular, how individuals of so many different nationalities came together in the plot.

What is IRT? As the TRAC report states, "IRT, also known as international extremism, is a transnational phenomenon. Its adherents generally overcome traditional national differences by concentrating on a common goal of achieving social change, under the banner of personal beliefs, through violence."[3]

The report also advises, "People are capable of hate, despair, and violence regardless of background, social standing or education."[4] It goes on to assert, "IRT adherents may not consider themselves to be citizens of any particular country, but instead seek common political, social, economic or personal objectives which transcend nation-state boundaries. The World Trade Center bombing provides an excellent example of this aspect of IRT."[5]

IRT seems to be a very dubious concept on its face. But to get an expert opinion, I called up Jim Fox and asked if he was familiar with IRT as an FBI acronym. He was not. I began to read from the report. He interrupted me, asking in disbelief, "Who wrote that?"

For a second expert opinion on IRT I called up Vince Cannistraro, former director of CIA Counterterrorism Operations. I read to him from the TRAC report. He had the same reaction as Fox. He interrupted, asking incredulously, "Is that in the report?"[6]

A State Department official who worked in the secretary of state's office read the section on IRT from the TRAC report and pronounced it "bizarre."[7]

Indeed, IRT explains everything and nothing. It suggests an incomplete, improperly conducted investigation. An investigation of a major bombing conspiracy should entail both a criminal investigation, aimed at catching perpetrators, and an intelligence investigation, aimed at explaining the big picture—which party, if any, was behind it.

The FBI's National Security Division was formally responsible for the intelligence investigation of the Trade Center bombing. It was the only agency that had both the evidence from the FBI investigation and the intelligence from the national security bureaucracies. And it clearly dropped the ball. It did not recognize the structure in which the individuals who carried out the bombing operated. Thus, its explanation of the bombing was reduced to one of violent individuals, and when it sought to elaborate on that analysis, it made little sense.

IRT in Other Forms

The Clinton administration's position on terrorism has been taken up and energetically promoted by a variety of individuals outside the government, the most prominent of whom is probably the journalist Steven Emerson. He produced the documentary film "Jihad in America," which aired on national television in November 1994.

Already in April 1993, little more than a month after the Trade Center bombing, Emerson wrote of the "frightening new brand of terrorism growing up on U.S. soil." The new terrorists were "not hit men dispatched from the Middle East" but rather had their roots in America. In fact, there were "hundreds of radical operatives" in the United States who "answer to no one but themselves." That was "the worst possible nightmare," because those networks were nearly impervious to penetration before they attacked, and there was no address for retaliation afterward. Emerson identified Mahmud Abu Halima as the "ringleader" in the Trade Center bombing, and suggested it was "the first . . . Egyptian fundamentalist-Palestinian terrorist operation."[8] That was IRT in op-ed format.

A year later, in producing "Jihad in America," Emerson employed as an adviser an Israeli expert on terrorism—someone with many years of government experience. In the fall of 1994, that adviser was directed to me. In early October, in two long sessions, we went through much of the material presented in this book.

The antiterrorism expert was stunned. He kept saying, "Stop. I have a headache. No, please go on. Give me a glass of water." He explained that he and Emerson had not gone through the evidence from the World Trade Center bombing trial as carefully as he and I were doing. Their basic purpose in "Jihad in America" was to show how dangerous the extremists were. They had used the Trade Center bombing only as "a hook" to make that point. And they were *not* saying that the bomb did *not* have state sponsorship.

I replied that that was fine, but they should say so explicitly; otherwise, people would think they were implying that the Trade Center bombing had no state sponsor. Subsequently, I spoke with Emerson and he told me the same thing—they were *not* saying that the bomb was *not* state-sponsored.

But when "Jihad in America" aired, it said nothing of the sort. It left the distinct impression that the bombing was the work of Muslim extremists, unaided by any state.[9] In subsequent publications Emerson conflated the two New York bombing conspiracies, thus ignoring the U.S. government's own position that none of those charged in the second bombing plot actually participated in the first. He explicitly stated that the Trade Center bombing "was carried out by an ad hoc coalition of affiliates of five different groups, including the Egyptian Al Gamaa al Islamiya, Hamas, Islamic Jihad, Al Fuqra, and the Sudanese National Islamic Front."[10] Later, Emerson boldly asserted that "the infrastructure [of the extremists] now exists to carry off twenty simultaneous World Trade Center bombings across the United States. And as chemical, biological, and even nuclear weapons become available to them, the threat becomes ever more ominous."[11]

Emerson had an important point. Muslim extremists were establishing themselves in America. They were engaged in deplorable political activities and in fundraising for terrorism, much of it conducted in countries allied to the United States. But Emerson went much further. He manipulated the evidence to make the extremist threat appear as terrifying as possible. In doing so, he contributed to obscuring the danger posed by state sponsors of terrorism, particularly Iraq. And he contributed to the problem he himself described—authorities can do virtually nothing about terrorism, when it is understood as he depicted it.

If Emerson was the most tenacious journalist flogging the Islamic threat, he was far from the only one. In January 1994, former CNN correspondent Peter Arnett screened a documentary entitled "Terror Nation? U.S. Creation?" It was about "blowback" from Afghanistan. U.S. support for the Afghan resistance to the Soviets had come back to haunt America as terrorism on U.S. soil, particularly in the Trade Center bombing. I asked Arnett how his view accounted for Abdul Rahman Yasin's presence in Baghdad. He did not even know there *was* a Trade Center bombing fugitive at large in Iraq.[12]

More recently, British journalist Simon Reeve, drawing heavily on official sources, published a book about Ramzi Yousef and Usama bin Ladin. In *The New Jackals* he asserts that "they epitomize the new terrorist threat facing the West," and he warns that "the West will soon be facing militants armed with weapons of mass destruction, and there is little that can be done to prevent a disaster."[13] That exceedingly pessimistic claim rests on the assumption that the "new terrorism" is carried out by elusive individuals, rather than by states. Yet, oddly enough, Reeve acknowledges that the World Trade Center bombing—often cited as the first example of, and the model for, the "new terrorism"— probably was *state-sponsored*: "The evidence of Iraqi involvement in the twin towers bombing is strong," Reeve writes.[14]

In spite of that concession to the evidence, Reeve nevertheless maintains that Iraq's suspected involvement in the Trade Center bombing was not after all a very significant matter. Reeve's claim is that Yousef acted as a terrorist for hire who committed many acts of terrorism in addition to the Trade Center bombing. According to Reeve, Iraq turned to Yousef to carry out that bombing attack just as others had used him to carry out other acts of terrorism.

According to Reeve, in addition to the plot to bomb U.S. airplanes, Yousef was behind a botched attempt to assassinate Pakistani prime minister Benazir Bhutto; the bombing of a major Shi'i shrine in Mashad, Iran; and a failed attempt to bomb an embassy in Thailand in March 1994.[15] Yousef also supposedly met with Oklahoma City bomber Terry Nichols in the Philippines, and there is a "wealth of circumstantial evidence" to suggest that Yousef was linked with the Oklahoma City bombing.[16] Iraq's involvement in the Trade Center bombing pales, apparently, when compared with all the other plots in which Yousef was involved. And the "evidence of Iraqi involvement in the rest of Yousef's crimes is virtually non-existent."[17]

But in fact the only plots in which Yousef was demonstrably involved were the Trade Center bombing and the conspiracy to bomb U.S. aircraft. Reeve's claim that Yousef was connected with other terrorist plots rests, in each case, on uncorroborated assertions and flimsy evidence. Moreover, Reeve assimilates Yousef into the loose network of Muslim extremists that supposedly represents the "new terrorism" in the face of significant evidence to the contrary, citing only thin, unverifiable information to make that claim.

As Reeve himself notes, Yousef is not even a pious Muslim, let alone a religious extremist. What, then, moved him to commit terrorism? Shortly after his arrest, Yousef asserted that his mother was Palestinian, but that claim remains uncorroborated. No one has determined the woman's name or where she is now, nor has anyone interviewed her.[18] Reeve nevertheless cites it as fact, explaining that Yousef's motive, at least in part, was the same as that of an earlier generation of radical Palestinian nationalists. For Yousef, according to Reeve, "Israel is an illegitimate state" and "the Palestinians have the right to attack Israeli targets and any other organization or country that interferes in support of Israel."[19] Reeve fails to address the question of why a Palestinian nationalist would deliberately allow—or plan—the arrest of the New York area extremists, including the Palestinian Salameh.[20]

Reeve cannot provide a coherent account of Yousef and his career as a "free-lance" terrorist, and thus fails to support his contention that Yousef's ties to Iraq were of a purely ad hoc character. The detailed information regarding Yousef is of course difficult to sort out and analyze. Moreover, the most important and reliable information about Yousef is the documentary evidence and transcripts of the trials in New York. One of Reeve's principal sources (acknowledged with "special thanks") was Neil Herman, recently retired from the FBI. Herman told me that his conversations with Reeve were exclusively by telephone from London. He was under the impression that Reeve had never actually come to New York on the case.[21]

Previous Experience of State Sponsorship

A fairly consistent pattern in the investigation of major Middle Eastern episodes of terrorism has been that while they may at first seem to be acts of individuals or shadowy groups, a state sponsor eventually emerges. For example, the taking of American hostages in Lebanon (beginning in 1984) at first seemed to be the work of a Lebanese Shi'i, Imad Mughniyah. His brother-in-law was being held in a Kuwaiti jail, along with sixteen others, for the 1983 car bombings of the U.S. and French embassies in Kuwait. Mughniyah repeatedly demanded that the "Dawa 17," as they were called, be released in exchange for the American hostages he held. Only over time—and with heated debates within the U.S. bureaucracies—did it become apparent that Iran was behind Mughniyah. Tehran had exploited his very real personal grievances for its own ends. Without Iranian backing, Mughniyah could not have carried off the hostage-taking. And Iran called the shots. When the project no longer served Tehran's purposes, the hostages were released.

Similarly, a wave of bombings and shootings directed against Jewish targets in Paris in 1985 and 1986 at first appeared to be the work of a shadowy, unknown Middle Eastern group. Again, with time, Iran proved to be behind that violence as well.

Elsewhere, too, states have proven to be far more involved in terrorism than was generally recognized at the time. Since the collapse of the Soviet Union, it has become much clearer that the Soviets and other Communist regimes provided support for terrorism carried out by others. After the Cold War's end, an archivist for the KGB, Vasily Mitrokhin, defected. Subsequently he wrote about the KGB's activities. As Mitrokhin explained, when Yuri Andropov became head of the KGB in 1967, he wanted the KGB to carry out "special operations"—terrorist actions—against the West. But Andropov hesitated, because of the possible consequences if the Soviets were caught red-handed. So he turned to proxies. Subsequently the Soviets provided assistance to the IRA and the Sandinistas. The latter, at Soviet behest in 1974, tried to kidnap the U.S. ambassador to Nicaragua, who only narrowly managed to escape.[22]

The KGB also recruited and aided Wadi Haddad, deputy chief of the Popular Front for the Liberation of Palestine. Haddad played a major role in exporting Palestinian terrorism to Europe in the 1970s and once attempted, on Soviet orders and assisted by Soviet intelligence, to capture the deputy CIA chief in Beirut.[23] And after Hungary's Communist regime fell, the new government revealed that Budapest had supported the international terrorist known as Carlos the Jackal.[24] Similarly, it is now recognized that East Germany and Hungary provided substantial assistance to the radical leftist Red Army Faction.

One can only underscore that the new theory of terrorism promulgated by the Clinton administration is at odds with America's previous experience with terrorism. Moreover, it is at odds with what would seem to be obvious—that the resources and experience of a terrorist state far exceed that of an individual, even if there may be advantages for both in working together.

Other Official Views

A blue ribbon presidential commission established to assess the future roles and capabilities of the U.S. intelligence community took issue with the administration's approach to dealing with terrorism. Its report, issued in 1996, noted that while law enforcement has an important role to play in fighting terrorism, "it may not be the most appropriate response in all circumstances."[25]

One of the problems is acquiring evidence. Criminal convictions in a U.S. court require proof beyond a reasonable doubt. That high standard has a purpose. It aims to protect the life, liberty, and property of U.S. citizens against abuse by authority. But that is not the standard that is used in national security affairs. It never could be, because such certainty rarely exists. In 1981, shortly after Ronald Reagan became president, the CIA was tasked with preparing a national intelligence estimate on Soviet support for terrorism. The draft report by agency analysts concluded that the Soviets did *not* support terrorism. It claimed that the Soviets "disapproved of terrorism, discouraged the killing of innocents by groups they trained and supported," and did not assist Third World terrorist organizations like Abu Nidal, or West European groups like the Red Army Faction. The analysts "cited Soviet public condemnations of such groups and carefully described the distinctions the Soviets made between national liberation groups or insurgencies and groups involved in out-and-out terrorism."[26]

Upon receiving that draft, CIA director William Casey protested that it had "the air of a lawyer's plea," arguing that there should be no indictment because "there was not enough evidence to prove the case beyond a reasonable doubt." But, as Casey noted, "the practical judgments on which policy is based in the real world do not require that standard of proof which is frequently just not available."[27]

Thus, as the 1996 presidential commission explained, "Compiling proof beyond a reasonable doubt—the standard in criminal cases—may be even more difficult with respect to global crime," including terrorism. And the commission suggested that "diplomatic, economic, military, or intelligence measures, in many cases, can offer advantages over a strict law enforcement response, or can be undertaken concurrently with law enforcement. . . . Under the Constitution, the President has responsibility not only to enforce the laws but also to conduct foreign policy and provide for the common defense."[28]

To deal effectively with terrorism, authorities need to be able to consider information that would not constitute evidence in a court of law. And if the terrorism involves Middle Easterners, authorities need to understand contemporary developments in that region in order to assess which parties are most likely to have been behind the attack. Yet such an understanding has been lost. The administration's incessant spin on Iraq has created the impression that Saddam has been so cowed by U.S. measures against him that he has been unable to take hostile action of his own. At the same time, the general understanding of terrorism has become confused, because the difference between the capabilities of individuals and states has become blurred.

That was not the prevailing approach in the 1980s. Following the 1983 bombing of the U.S. marine barracks in Beirut, for example, no one was ever arrested. Nor did any decisive evidence emerge to tie the attack to a state sponsor. Nevertheless, Syria and Iran were assumed to be behind the bombing, because their avowed intent was to drive the United States from Lebanon. If today's understanding of terrorism had prevailed then, could such a conclusion ever have been reached? Why couldn't the bombing have been the work of individuals seeking to get the United States out of Lebanon, or of a "loose network" that had come into existence during Lebanon's chaotic civil war?

In many terrorist incidents, the evidence available at the scene of the crime is likely to be inconclusive. It may not suffice to allow authorities to determine the individuals involved in the attack, let alone those who sponsored them. Moreover, any relevant intelligence, such as a threat environment, is likely to be only suggestive. Still, it is vitally important to be able to conduct an informed discussion about which parties are most likely to have been responsible. For authorities to fail to offer any reasoned explanation of the origin of a particular terrorist attack means simply that those behind the terror got away with it, and it increases the likelihood that they will attack again. The consequences could be truly catastrophic if the terrorism should become unconventional.

Indeed, an advisory panel mandated by Congress took issue with the administration's view of who might carry out a devastating unconventional terrorist attack. The administration claims that the "new terrorists" are capable of carrying out such an attack. Therefore the likelihood of such an attack is high, and it is necessary to take extensive civil defense measures to be able to deal as ably as possible with the consequences of such an attack.

But the advisory panel, headed by Virginia governor James Gilmore, thought otherwise. It reviewed the very considerable technical and organizational obstacles to carrying out such terrorism. The panel concluded that while some terrorists, including radical religious organizations or sects, might be inclined to undertake terrorism using unconventional weapons, "they lack the means to translate their desire for mass murder into effective action."[29] In effect, only a state has the capability to carry out such terrorism.

Therefore, the danger would really exist only if a terrorist state assisted a terrorist group. Then "the prospect for a true act of mass destruction would become a distinct possibility."[30] That indeed is a scenario that Clinton warned about as the second Iraq crisis began in January 1998.[31]

Nevertheless, the panel reasoned that that was not likely to happen. States would be "unlikely to want to place" such dangerous weapons "in the hands of groups over which they have no ultimate control." And they would be so concerned about the response, if their role in such terrorism were discovered, that they "would likely be deterred from considering this course of action."[32]

Yet the one case in which the panel suggested a major unconventional terrorist attack could happen—a state assisting others—has already occurred. That was the World Trade Center bombing, intended to topple New York's tallest tower onto its twin and, according to New York authorities' suspicions, to disperse cyanide gas. But the attack was not generally recognized for what it was—an Iraqi "false flag" operation. Why, then, should Saddam believe that an unconventional terrorist attack would be treated differently—particularly a biological attack, in which those who carried it out would have fled even before authorities recognized an attack had occurred? Can we be confident that, given the Clinton administration's handling of terrorism and the broader misunderstandings that it has encouraged, Saddam would be deterred from carrying out such an attack?

20

Conclusion

Some years ago a State Department official spoke ruefully of "the train wreck coming." [1] He was frustrated with the Clinton administration's treatment of terrorism as a law enforcement issue and the FBI's refusal to share information about its investigations. Six years later, the train wreck has arrived.

There were plenty of warning lights flashing along the way. About once a year, on average, starting with the 1993 attack on the World Trade Center, a major bombing occurred or a terrorist conspiracy was thwarted. And each time a plot was uncovered, the "loose networks" of militant Muslims provided a ready explanation—particularly the ubiquitous bin Ladin network.

It seemed incredible even then that one man could have accomplished everything that was attributed to him.

If we could recover the understanding of terrorism that we had a decade ago, we would recognize that a state is behind the September 11 assault and the biological attacks that have followed. And we would ask, Which state could it be? The only reasonable answer is Iraq, with which we are still at war. After all, we bomb Iraq on a regular basis and maintain an economic siege that is itself an outgrowth of the Gulf War. It is only to be expected that Saddam would take action against us. (This does not mean, however, that if we lifted sanctions and withdrew from the Gulf, Saddam would forgive and forget; such a show of weakness would only encourage further attacks.)

Over time, we will learn more details about the September 11 attacks. But can we afford to wait for the "full" story before we protect ourselves from further attacks? America is at war, and our country's leadership must shift, very decisively, from the relatively leisurely pace of peacetime deliberations. In war, decisions must be made on the basis of incomplete information—before it becomes too late. In short, decisions must be made on the basis of *probabilities*.

What is the probability of a future attack? By all accounts, it is very high. What is the probability that bin Ladin carried out the September 11 assault without assistance from a state? Next to zero. How will U.S. action—or inaction—affect the likelihood of future attacks? If the United States fails to assert that Iraq was involved and to take the appropriate measures, Saddam will have little reason to desist. That dynamic may already be in play. Even after America launched its war on bin Ladin, terrorism has continued in the form of the anthrax letters.

We already have the clues to show us that Iraq is probably involved. Bin Ladin has known ties to Iraqi intelligence. Bin Ladin's aims, moreover, coincide with Iraq's agenda: to overthrow the Saudi government, to end the U.S. presence in the Gulf, and to have the sanctions on Iraq lifted. In Baghdad, where any mass protest must be officially sanctioned, Iraqis have marched in support of bin Ladin. *Babil,* a newspaper owned by Saddam's son Uday, has repeatedly praised bin Ladin.

The Iraqi press has threatened further attacks against the United States. *Babil* even threatened that biological terrorism would follow the U.S. intervention in Afghanistan:

> At this stage it is possible to turn to biological attack, where a small can, not bigger than the size of the hand, can be used to release viruses that affect everything. This attack might not necessarily be launched by the Islamists. It might be done by the Zionists or any other party through an agent. The viruses easily spread by air, and people are affected without feeling it.[2]

Saddam is the only world leader who publicly praised the September 11 attacks and who suggested that they were justified. Indeed, Saddam—a man known to have used chemical weapons against dissenting regions of his own country—proclaimed the following day,

Regardless of the conflicting human feelings about what happened in the United States yesterday, the United States reaps the thorns that its rulers have planted in the world. . . . Those who consider the lives of their people as precious and dear must remember that the lives of people in the world are also precious and dear to their families.

As we have seen in this book, the evidence from the 1993 World Trade Center bombing points strongly to Iraq:

• Mohammed Salameh, convicted for his role in the Trade Center bombing, made forty-six calls to Iraq, the majority to his uncle in Baghdad, a convicted terrorist who spent eighteen years in an Israeli jail. Most telephone conversations in Iraq are monitored by Iraqi intelligence.

• Following Salameh's calls to his uncle, Abdul Rahman Yasin, an Iraqi resident in Baghdad, traveled to Jordan, where he obtained a U.S. passport. One of the original indicted conspirators, Yasin returned to Iraq after the Trade Center bombing and has been harbored by Iraq ever since.

• Ramzi Yousef entered the plot only after Salameh's calls to his uncle. He transformed the conspiracy from a pipe bombing plot to the audacious attack on the Trade Center by using the "largest improvised explosive device in the history of forensic explosives."

• Someone created a false identity for Yousef by tampering with the official file of Abdul Basit Karim, who lived in Kuwait. That included substituting a fingerprint card with Yousef's prints on it for the original card. Reasonably, only Iraq could have done that, while it occupied Kuwait.

• Sudanese intelligence conspired with another group of Islamic extremists in the related plot against the United Nations and other New York sites, at a time when the Sudanese cooperated closely with Iraqi intelligence.

All of this evidence adds up to the very substantial probability that Iraq was involved in both of the World Trade Center
attacks. The Islamic extremists provide Iraq with an ideal
smokescreen for its own publicly declared terrorist aims. As long
ago as 1998, an Arab intelligence officer acquainted with
Saddam predicted "large-scale terrorist activity run by the
Iraqis, conducted under 'false flags'"—specifically under the
cover of bin Ladin's organization.

Furthermore, the U.S. government consistently failed to act
on what it knew, virtually ensuring that Iraq would avoid the
consequences of promoting terrorism:

• Abdul Rahman Yasin was picked up by the FBI and subsequently released as a cooperative informant—even though he
had actually assisted in mixing the chemicals for the bomb. He
immediately traveled to Baghdad, where he remains. *Only after
the September 11 assault—more than eight years later—has
Washington finally made an issue of Yasin's presence in
Baghdad.*
• In August 1993, Shaykh Omar was indicted on conspiracy
charges in the second New York bombing plot, targeting the
United Nations, New York's Federal Building, and the Lincoln
and Holland tunnels. *The U.S. government had specific information that a state was involved in this massive plot.* Yet that evidence was only acknowledged three years later, when the
Clinton administration finally revealed that two Sudanese
diplomats had been directly involved. Moreover, given the close
ties between the Sudanese and Iraqi governments (both of them
Arab and Sunni Muslim), one obvious possibility—never officially acknowledged—was that Sudan was acting on behalf of
Iraq.
• The trial for the African embassy bombings made clear that
al-Qaeda and Sudanese intelligence were intimately connected.
Sudanese intelligence agents had approached bin Ladin in
Afghanistan in 1991 and invited him to come to Khartoum. In
Khartoum, Iraqi intelligence established its own ties to bin
Ladin, as the *Wall Street Journal* reported, and in December
1998, a senior Iraqi intelligence official, Farouk al-Hijazi, visited
bin Ladin in Afghanistan. *Far from operating as a hermit camped
in the remote vastness of the Hindu Kush, bin Ladin has served*

as a link between individual militants and terrorist states from the beginning of his operation.

- *U.S. intelligence agencies have shown little interest in gleaning information from Iraqi defectors,* even in the case of a prominent nuclear scientist. The *Washington Post*'s Jim Hoagland cites the case of a former Iraqi intelligence officer who had observed suspicious training activity on a Boeing 707 a year before the September 11 attacks. Even a month after the attacks, "[CIA officers in Ankara] reportedly showed no interest in pursuing a possible Iraq connection to September 11."[3]

- In June and July of 2001, the CIA picked up a series of intercepted communications that strongly suggested bin Ladin was planning a major operation, and then issued a number of alerts about possible terrorist attacks on U.S. targets overseas that never materialized. In contrast, the September 11 assault came without any warning—perhaps the biggest U.S. intelligence failure since Pearl Harbor. How is it that U.S. intelligence could so easily pick up the warnings about some of bin Ladin's plotting but completely miss the most massive terrorist assault in history? The people organizing these attacks are evidently quite capable of maintaining the security of their communications, especially given our reliance on technical (rather than human) means of intelligence gathering. The vast bulk of their communications that the United States intercepts is either insignificant or meant to fall into U.S. hands. And the intercepted communications serve to sustain the illusion that al-Qaeda alone is planning major attacks and capable of carrying them out.

The CIA has, for some years, been prominent among the bureaucracies in Washington that promote the view that bin Ladin is the biggest terrorist threat to America. The current CIA director, George Tenet, headed the intelligence desk in the Clinton White House before being appointed Director of Central Intelligence in 1997. He was closely associated with the theory of "loose networks" that developed on Clinton's watch. As recently as February 7, 2001, Tenet testified before Congress,

State-sponsored terrorism appears to have declined over the past five years, but transnational groups with decentralized leadership that makes them harder to identify and disrupt are emerging. We are seeing fewer centrally controlled operations, and more acts initiated and executed at lower levels. . . . Usama bin Ladin and his global network of lieutenants and associates remain the most immediate and serious threat.

As we have shown in *The War Against America,* it is time to reassess this understanding, and to recognize that the global network extends not only to individuals, but to sponsor states and governments.

The exclusive focus on figures like bin Ladin and Shaykh Omar has had far-reaching strategic consequences. Saddam himself—never known for religious observances—makes an extremely unconvincing focus for a pan-Islamic jihad. Yet Usama bin Ladin (like Shaykh Omar) presents a very different sort of political challenge. In the present crisis, the Taliban network has succeeded in mobilizing supporters as far away as Indonesia, under the banner of Muslim solidarity. If in fact the al-Qaeda network is acting as the arm of an Iraqi terrorist campaign, it is a strategic approach designed to effect a stunningly ambitious realignment of global political alliances. It transforms what might have been an American war with Iraq into a far more amorphous conflict, which some will inevitably view as a war with Islam.

The symbolism of the September attacks was so striking, and their psychological effect so shattering, that it has been easy to think of them purely as acts of terrorism. Once we understand them primarily as acts of war, however, we will have come closer to understanding the situation we confront. The World Trade Center was not just one of the world's tallest structures, not just the "arrogant" symbol of American commerce, not just a hugely concentrated civilian population. In the simplistic understanding of our enemy, the World Trade Center appeared as a vital center of the U.S. economy—just as the Pentagon represented the nerve center of our military organization. These attacks were intended, presumably, as a decisive first strike that would cripple our capacity to wage war, just as the attack on Pearl Harbor was designed to destroy our maritime capability.

The attacks of September 11 failed in this regard, owing on the one hand to the complexity of our real-world economy, and on the other to the miraculously well-timed structural renovations that had been completed at the Pentagon. (Ample credit is also due to the truly heroic and successful efforts of the passengers who took their plane into the ground rather than permitting it to strike its intended target—very possibly the Pentagon.) But the attacks were not just a wake-up call, or an expression of virulent hatred, or a particularly emphatic commentary on globalization. They were a calculated move in a war that we are only now beginning to recognize as our generation's defining challenge.

Finally, a very sobering question remains to be addressed. Could the events of September 11 have been avoided? More precisely, were there policy decisions that might have been taken to address the terrorist threat in its early stages? Historians will certainly debate that issue. The motive behind the research and writing of this book was, very simply, my belief that such policies were urgently needed in order to forestall another murderous attempt, at some unknown time and place, to realize the horrifying ambition behind the 1993 bombing of the New York World Trade Center. It is especially disturbing that the next massive assault, when it came, targeted the very same place. One is left with the grim possibility that if the 1993 bombing of the World Trade Center had been properly investigated and the question of state sponsorship had been properly addressed, those who perished on September 11 would be alive today.

Appendix A

ABDUL BASIT'S 1984 PASSPORT

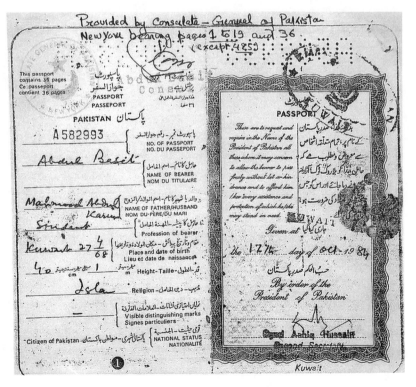

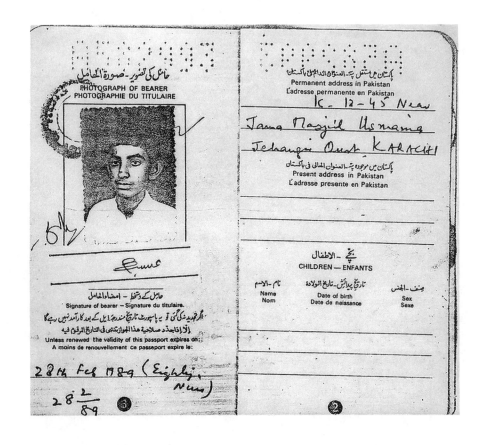

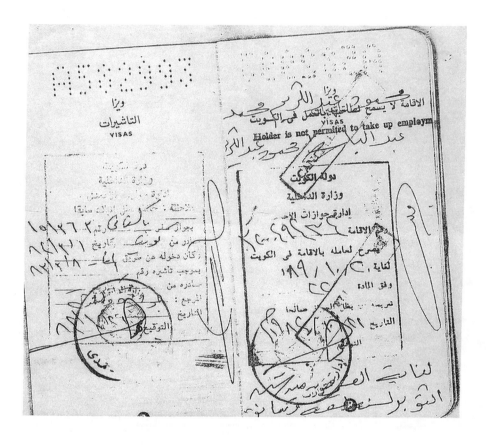

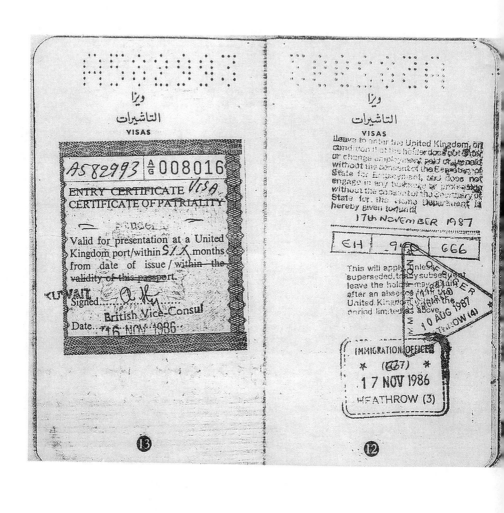

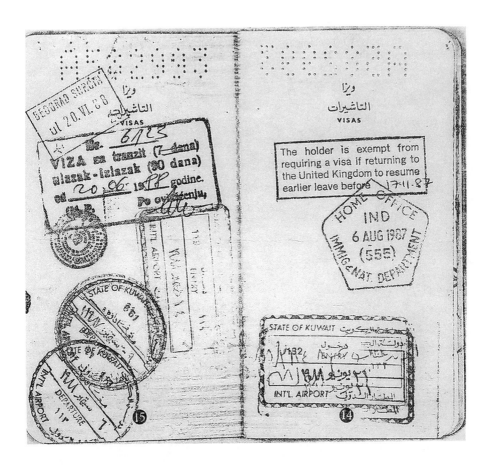

SOURCE: Government Exhibit 739 C, from the World Trade Center bombing trial. Office of the U.S. Attorney, Southern District of New York. Photograph by Daniel Brinzac.

Appendix B

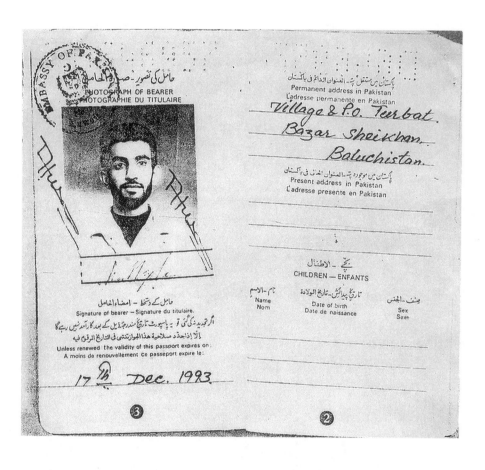

وہ ممالک جن کے لیے یہ پاسپورٹ کار آمد ہے
البلاد التي لها تكون صلاحية هذا الجواز

COUNTRIES FOR WHICH THIS PASSPORT IS VALII
PAYS POUR LESQUELS CE PASSEPORT EST VALABL

Afghanistan, Algeria, Angola, Argentina, Australia,
Austria, Bahamas, Bahrain, Barbados, Belgium,
Bangladesh, Benin, Bolivia, Botswana, Brazil, Burma,
Burundi, Cameroon, Canada, Central African Republic,
Chad, Chile, China, Colombia, Comoros, Congo, Costa
Rica, Cyprus, Denmark, Djibouti, Dominican Republic,
Ecuador, Egypt, El-Salvador, Ethiopia, Equatorial
Guinea, Federal Republic of Germany, Fiji, Finland,
France, Gabon, Gambia, Ghana, Greece, Grenada,
Guatemala, Guinea, Guinea Bissau, Guyana, Haiti,
Honduras, Iceland, Indonesia, Iran, Iraq, Ireland, Italy,
Ivory Coast, Jamaica, Japan, Jordan, Kampuchea,
Kenya, Kuwait, Lebanon, Lesotho, Liberia, Libya,
Luxembourg, Madagascar, Malawi, Malaysia,
Maldives, Mali, Malta, Mauritania, Mauritius, Mexico,
Morocco, Mozambique, Nepal, Netherlands, New
Zealand, Nicaragua, Niger, Nigeria, Norway, Panama,
Papua New Guinea, Paraguay, People's Democratic
Republic of Yemen, Peru, Philippines, Portugal,
Qatar, Rwanda, Sao Tome & Principe, Saudi Arabia,
Senegal, Seychelles, Sierra Leone, Singapore,
Somalia, Spain, Sri Lanka, Sudan, Sultanate of
Oman, Surinam, Swaziland, Sweden, Switzerland,
Syria, Tanzania, Thailand, Togo, Trinidad and
Tobago, Tunisia, Turkey, Uganda, United Arab
Emirates, United Kingdom, United States of America,
Upper Volta, Uruguay, Venezuela, Yemen Arab
Republic, Zaire, Zambia, Zimbabwe,
Also Valid for Hong Kong.

⑤

تجديدات
التجديدات

RENEWALS
RENOUVELLEMENTS

④

راحظات
ملحوظات
OBSERVATIONS

The holder of the Passport
previously travelled on Passport

No. *A582993*

dated   *17. 10. 1984*

Issued :   *Kuwait*
which have been cancelled/reported
lost/is attached bearing valid visa
and returned to the holder .

OBSERVATIONS

KUWAIT

OBSERVATION

No. 1865/10 Date 22.5.10
PROFESSION HAS BEEN CHANGED
At Page ....1........ To read as

"Electronic Engineering"

Mohammad Akhtar Khan,
Second Secretary,
Embassy of Pakistan,
Kuwait.

7

6

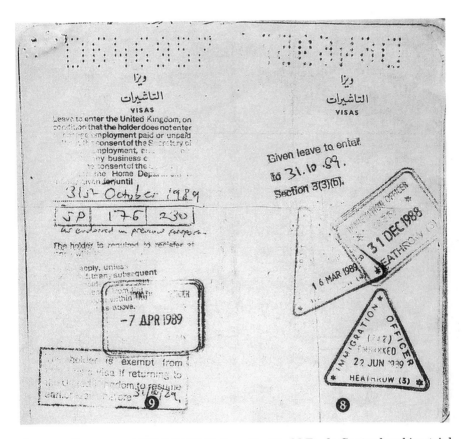

SOURCE: Government Exhibit 739 B, from the World Trade Center bombing trial. Office of the U.S. Attorney, Southern District of New York. Photograph by Daniel Brinzac.

WHAT RAMZI YOUSEF WITHHELD—AUTHOR'S SUPPOSITION

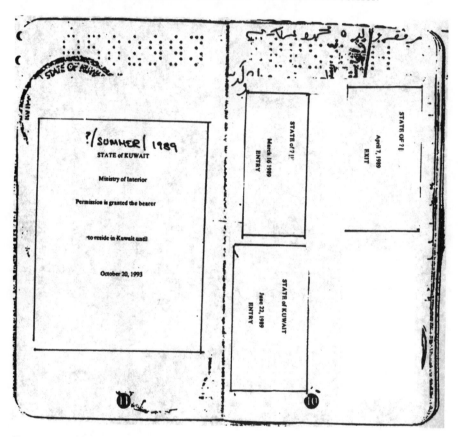

SOURCE: Author.

Notes

Chapter 1: Introduction

1. See, for example, Steven Emerson, "A Terrorist Network in America?" *New York Times*, April 7, 1993; most recently, U.S. Department of State, *Patterns of Global Terrorism, 2000* (Washington, D.C., Department of State Publication, April 2001).

2. Jim Fox first told this author of his belief that Iraq had been behind the Trade Center bombing in October 1994. Our discussions continued until shortly before his death in May 1997.

3. This conversation occurred at a meeting hosted by the Office of the District Attorney of New York, January 13, 1995.

4. Pentagon official, to author, March 1995.

5. State Department official, to author, March 1995.

Chapter 2: Assassination of Meir Kahane

1. *Sunday Star-Ledger*, December 2, 1990; *New York Newsday,* November 7 and 19, 1990.

2. *New York Times*, December 9, 1991.

3. *Village Voice*, November 12, 1991.

4. William M. Kunstler, with Sheila Isenberg, *My Life as a Radical Lawyer* (New York: Birch Lane Press, 1994), p. 320.

5. *New York Times*, December 6, 1990.

6. *New York Post*, December 19, 1990; *Newsday*, December 19, 1990.

7. *New York Newsday*, November 4, 1991.

8. *National Law Journal*, December 30, 1991; *New York Times*, December 9, 1991.

9. *New York Times*, November 25, 1991.

10. *New York Newsday*, December 23, 1991.

11. *New York Times*, January 30, 1992.

12. Ibid., December 23, 1991.

13. *Washington Post*, December 23, 1991.

14. *New York Newsday*, November 8, 1990.

15. *New York Times*, November 5, 1991.

16. *New York Post*, December 17, 1991.

17. Yossi Melman, *The Master Terrorist: The True Story behind Abu Nidal* (New York: Adama Books, 1986), p. 85.

18. *New York Newsday*, December 8, 1990.

19. *Jerusalem Post*, October 28, 1990.

20. Agence France-Presse, November 1, 1990, in Foreign Broadcast Information Service (FBIS), Near East and South Asia, November 2.

21. *Wall Street Journal*, June 16, 1993.

22. An exception was Robert I. Friedman, in his article "Behind the Kahane Killing: Voice Investigation Reveals Links to Arab Terror Groups," *Village Voice*, November 5, 1991.

CHAPTER 3: ORIGINS OF THE WTC BOMBING

1. *New York Times*, January 19, 1994.

2. Vincent Cannistraro, to author, October 12, 1994.

3. *New York Times*, March 6, 1995.

4. *New York Newsday*, June 30, 1992.

5. Government's Memorandum of Law in Opposition to Defendants' Pretrial Motions (Phase 1), p. 17, United States v. Omar Abdel Rahman et al.

6. Ibid., p. 18; Taped discussion between Emad Salem and Siddig Ali, May 31, 1993; *New York Times*, March 14, 1995.

7. Government's Memorandum of Law in Opposition to Defendants' Pretrial Motions (Phase 1), p. 18, United States v. Omar Abdel Rahman et al.

8. Ibid., p. 20.

9. *New York Times*, March 3, 1995.

10. Ibid., November 1, 1993.

11. Transcription of Salem's undated telephone calls: Source Tape no. 19, p. 10. Emad Salem regularly taped conversations on his home

telephone. But the FBI did not know that. It only emerged in June 1993, when the second New York bombing investigation concluded, as authorities sought to move him out of his apartment, for his own safety. Once the tapes were discovered, Salem did not want to turn them over, but the FBI insisted it was necessary.

12. Vincent Cannistraro, to author, October 12, 1994.

13. Those figures actually understate the change. Salameh's phone service was cut off July 9. The vast majority of Salameh's calls were made in June. On June 23, MCI cut off Salemeh's service and he was obliged to use an AT&T calling card for long distance calls. MCI billed the calls to Iraq in June, and it charged calls made after June 10 to all other countries to Salameh's July bill, while AT&T billed the bulk of the card calls in July.

14. *New York Times*, February 15, 1994.

CHAPTER 4: THE ACCUSED

1. *New York Times*, March 8, 1993.

2. *Los Angeles Times*, March 9, 1993.

3. Christopher Dickey, to author, during ABC/*Newsweek* investigation of Trade Center bombing for which the author was consultant, April–June 1994.

4. Judge Kevin Duffy, sentencing hearing of the Trade Center bombers, May 24, 1994.

5. *Time*, October 4, 1993.

6. *New York Times*, March 25, 1993.

7. That was captured by Steven Emerson, who described Abu Halima as "the alleged ringleader" in "A Terrorist Network in America?" *New York Times*, April 7, 1993.

8. *Newsday*, June 15, 1993.

9. *Newsday*, July 16, 1993.

10. *New York Times*, January 28, 1996.

11. Jim Fox, on "America's Most Wanted" segment on "Day 1," ABC News, June 27, 1994.

12. Eric Pilker, affidavit, March 5, 1993.

CHAPTER 5: MYSTERIOUS MASTERMIND

1. *Newsweek*, July 4, 1994.

2. Ajaj telephone conversation, September 12, 1992, taped while he

was in jail, Government Exhibit 662-A, pp. 11, 12, United States v. Muhammad Salameh et al.

3. Richard Bernstein, "Questions Raised in One Conviction in Blast at Towers," *New York Times*, October 17, 1994. The author is indebted to Mr. Bernstein for sharing details of his work.

4. Jamal Kashoggi, to author, January 1994.

5. *New York Times*, February 21, 1995.

6. Ajaj telephone conversation, September 12, 1992, Government Exhibit 662-A, p. 12; Ajaj telephone conversation, September 13, 1992, Government Exhibit 663, p. 9, United States v. Muhammad Salameh et al.

7. Ajaj telephone conversation, September 13, 1992, Government Exhibit 663, United States v. Muhammad Salameh et al., p. 11.

8. Sheila MacVicar, in conversation with the author, during the ABC/*Newsweek* investigation, April–June 1994.

9. Justice Department official, to author, April 1995.

10. Tony Geraghty, memo to David Marash and Laurie Mylroie, January 16, 1995.

11. *The Merck Manual*, 15th edition (Rahway, N.J.: Sharp and Dohme Research Laboratories, 1987), chapter 19, p. 153.

12. Dr. Robert Ratner, to author, summer 1994.

13. Agence France-Presse, August 18, 1990, in FBIS, August 21, 1990.

14. Officials at Swansea Institute, to author, February 21, 1996.

15. Tariq Mirza, to author, September 15, 1994.

16. Marvin Smilon, to author, September 1994.

17. Official, Bureau of Diplomatic Security, U.S. Department of State, to author, September 1994.

18. Jim Fox, to author, September 15, 1994.

19. Itamar Rabinovich, to author, December 15, 1994.

20. A Kuwaiti official read the contents of the Kuwaiti file on Abdul Basit Karim to the author, November 18, 1994.

CHAPTER 6: THE ESCAPE ROUTE

1. Kamran Khan, "CIA Slaying Suspect Sighted in Pakistan: Authorities Say Man May Have Fled to Iran," *Washington Post*, February 12, 1993.

2. Selig Harrison, *In Afghanistan's Shadow: Baluch Nationalism and Soviet Temptations* (Washington, D.C.: Carnegie, 1981), p. 7.

3. Mir Ahmad Khan Baluch, *Inside Baluchistan, a Political Autobiography* (Karachi: Royal Book Company, 1975), p. 58.

4. Historians discount this legend. They believe that the Baluch came from the north. But even if the Baluch came from Aleppo, Aleppo was not Arab at the time of their migration, several millennia ago.

5. Harrison, *In Afghanistan's Shadow*, p. 106.

6. Some contend that Pakistani authorities really found the weapons in the Iraqi consulate in Karachi and took them to Islamabad, as revealing them in the capital would have greater impact.

7. Harrison, *In Afghanistan's Shadow*, p. 37.

8. Ibid., p. 108.

9. Ahmed Mowafiq Zaydan, "Iraqi Oil Smuggled in Large Quantities to Pakistan and Iran," *Al Hayat*, December 16, 1993.

10. *New York Times*, February 13, 1995.

11. Harrison, *In Afghanistan's Shadow*, p. 8.

12. Ibid., p. 87.

13. Government's Memorandum of Law in Opposition to Defendants' Pretrial Motions (Phase 1), p. 24, United States v. Omar Abdel Rahman et al.

CHAPTER 7: THE BOMB ITSELF

1. *New York Times*, October 23, 1997.

2. *New York Newsday*, March 1, 1993.

3. Jim Fox to ABC News, reported to author by Sheila MacVicar, during ABC/*Newsweek* investigation, April–June 1994.

4. Jim Fox, to author, April 19, 1996.

5. Judge Kevin Duffy, sentencing hearing of the Trade Center bombers, May 24, 1994.

6. *New York Times*, January 21, 1994.

7. Memo to ABC/*Newsweek* investigation, May 8, 1994. The reporter also explained that U.S. authorities had provided the Israelis with little information on the Trade Center bombing.

8. *Financial Times*, July 3, 1991; *Nightline*, "Saddam's Chemical Connection," July 2, 1991.

9. *Financial Times*, July 3, 1991.

10. Louis Champon, to author, August 1, 1994.

11. An Iraqi defector who worked in one of Hussein Kamil's plants in this period, to author, July 1992.

12. Kenneth Timmerman, *The Death Lobby: How the West Armed Iraq* (Boston: Houghton Mifflin Company, 1991), p. 273.

13. Ibid.

14. *Nightline*, "Saddam's Chemical Connection," July 2, 1991.

15. This occurred at a presentation made by the author in December

1994 on the Trade Center bombing, to individuals in the U.S. intelligence community.

16. William S. Cohen, "Preparing for a Grave New World," *Washington Post,* July 26, 1999.

CHAPTER 8: STRUCTURE OF THE CONSPIRACY

1. *Village Voice,* March 30, 1993.

2. Statement of Sherman M. Funk, inspector general of the Department of State and the Arms Control and Disarmament Agency, before the Committee on Foreign Affairs, Subcommittee on International Security, International Organizations and Human Rights, House of Representatives, July 22, 1993, p. 21.

3. Most recently, a British journalist, Simon Reeve, claimed that the CIA arranged Shaykh Omar's U.S. visa in an effort to befriend him, in advance of the possible overthrow of the Egyptian government. *The New Jackals: Ramzi Yousef, Osama bin Laden, and the Future of Terrorism* (Boston: Northeastern University Press, 1999), p. 60.

4. Shaykh Omar, however, was not the "spiritual leader" of the assassins. A twenty-seven-year-old engineer, Abd al-Salam Faraj, had that role. He articulated the rationale for Sadat's assassination in a clandestinely printed book titled *The Absent Precept.* After the assassination, Faraj was tried and executed. See Emmanuel Sivan, *Radical Islam: Medieval Theology and Modern Politics* (New Haven: Yale University Press, 1985), p. 103. But Faraj lacked the qualifications to issue a formal religious ruling. Shaykh Omar, as a graduate of al-Azhar, Egypt's leading institution for the study of Islam, did have those qualifications, which is why the conspirators turned to him to obtain formal approval to assassinate Sadat.

5. *New York Newsday,* June 10, 1993.

6. *New York Times,* March 14, 1994.

7. Jim Fox, to author, November 6, 1996.

8. See Richard Bernstein, "Questions Raised in One Conviction in Blast at Towers," *New York Times,* October 17, 1994, for another report suggesting Ajaj's innocence.

9. *New York Newsday,* May 6, 1993.

10. The government accidentally erased four or five taped conversations between Ajaj and Yousef. However, there is no reason to believe that they were significantly different from the other conversations, and, as the defense pointed out, it would be no more legitimate for the government to claim that those tapes contained the definitive evidence

against Ajaj than for the defense to claim they contained the definitive evidence exculpating him.

11. Ahmad Ajaj, sentencing hearing of the Trade Center bombers, May 24, 1994.

12. Ajaj telephone conversation, December 29, 1992, Government Exhibit 674-A, p. 22, United States v. Muhammad Salameh et al.

13. Ajaj telephone conversation, September 12, 1992, Government Exhibit 662-A, p. 5, United States v. Muhammad Salameh et al.

14. United States v. Muhammad Salameh et al., p. 8395.

15. Ajaj telephone conversation, September 12, 1992, Government Exhibit 662-A, pp. 6, 10, United States v. Muhammad Salameh et al. About jail, Ajaj said, "It's a first class hotel. . . . Their treatment . . . towards us is very good. . . . I never had similar treatment anywhere else in the world."

16. Ajaj telephone conversation, October 7, 1992, Government Exhibit 667-A, p. 8, United States v. Muhammad Salameh et al.

17. Ajaj telephone conversation, December 29, 1992, Government Exhibit 674-A, p. 11, United States v. Muhammad Salameh et al.

18. FD-302, of Nidal Ayyad, July 8, 1993. "FD-302" is the FBI designation for a narrative, investigative form that records interviews and observations an agent makes. In this case it refers to Ayyad's proffer—a statement made by a defendant to law enforcement officials, revealing what he knows about a crime to see if authorities are willing to reach an agreement with him to reduce the charges against him in exchange for his cooperation. None of the information a defendant provides in a proffer can be used against him.

19. A poll conducted by the Trends Research Institute reported that 97 percent of Americans would not have returned for a $400 deposit if they had been involved in the World Trade Center bombing. Ninety-one percent said they would not return for all the money in the world; 6 percent said they would return for one million dollars or more; 2 percent said they would go back for $400. *New York Newsday*, April 15, 1993.

20. Former UNSCOM chairman Rolf Ekeus explained that UNSCOM once came across an Iraqi manual for terrorism and assassination. One of the instructions was that when acts of terrorism were undertaken, they should be made to look as if Iran were responsible. Ekeus, to author, May 22, 1997.

Similarly, after the November 13, 1995, bombing of a U.S. facility in Saudi Arabia, Hussein Kamil attributed the attack to Iraq, remarking that Saddam "wants everybody to think it's the Iranians behind the attack. He is using them as a smokescreen, hoping to fool the world." Uri Dan and Dennis Eisenberg, "Sly Snake of Baghdad," *Jerusalem Post*, November 23, 1995.

21. Ze'ev Schiff and Ehud Yaari, *Israel's Lebanon War* (New York: Simon & Schuster, 1984), pp. 100 ff.

CHAPTER 9: QUESTION OF STATE SPONSORSHIP

1. Transcription of Salem's undated telephone calls: Source Tape, No. 31, p. 9.

2. Source Tape, No. 37, p. 7.

3. Statement of Abdul Rahman Haggag to New York law enforcement, Government Exhibit 35119-E, United States v. Omar Abdel Rahman et al.

4. FD-302, of Abdul Rahman Haggag, on August 11–12, 1994, p. 3.

5. FD-302, of Nidal Ayyad, on July 8, 1993, pp. 4, 6.

6. Ibid.

7. The CIA was slow to recognize this fact. CIA director James Woolsey testified before the Senate Intelligence Committee on January 25, 1994. He claimed that the World Trade Center bombing was the work of a combination of local Sunni and Shi'i fundamentalists.

8. Ralph Blumenthal, "Missing Bombing Case Figure Reported to Be Staying in Iraq," *New York Times,* June 10, 1993.

9. Jim Fox, to author, December 1, 1994.

10. Gil Childers, to author, December 15, 1995.

11. That was also the view of Massoud Barzani, who heads the Kurdistan Democratic Party. Barzani is the son of the legendary Kurdish leader Mullah Mustafa Barzani, and he was his father's intelligence officer. He has a good understanding of Saddam's regime. The author discussed Musab's phone calls to his brother in Baghdad with Barzani on July 12, 1993. Not only did he agree, but he was amused by what the FBI had done. He considered it an example of American naïveté in dealing with Iraq.

12. This information comes from the author's personal conversation on November 21, 1994, with an Iraqi relation of Abdul Rahman who lived in Great Britain.

13. During a speech he gave in Buenos Aires, Argentina, Freeh explained that all the indicted World Trade Center fugitives had been tried and convicted, save for "Abdul Rahman Yasin [who] is still at large, hiding in his native Iraq"; May 12, 1998.

14. Allan Gerson, to author, spring 1995.

15. *New York Times*, August 1, 1994.

16. Taha Yasin Ramadan, then first deputy prime minister, Baghdad Domestic Service, February 15, 1991, in FBIS, February 19.

17. Baghdad Domestic Service, February 6, 1991, in FBIS, February 6.

18. Iraqi News Agency, January 30, 1991, in FBIS, January 30.

19. The date of the Trade Center bombing was Friday, February 26, 1993. Thus, the precise anniversary of the cease-fire fell on Sunday. Given that the conspirators' intent was to cause maximal casualties, they would not have carried out the bombing on the weekend. Alternatively, February 26 was also the date of Kuwait's liberation.

20. Ron Kessler, *The FBI* (New York: Simon & Schuster, 1993), p. 37.

21. Terrorism Research and Analytical Center, Counterterrorism Section, Intelligence Division, FBI, "Terrorism in the United States: 1991," p. 20.

22. Scott Ritter, *Endgame: Solving the Iraq Problem—Once and For All* (New York: Simon & Schuster, 1999), p. 121.

23. "The View From Baghdad: Tariq Aziz," *Middle East Quarterly*, June 1994, pp. 64–65.

24. Baghdad Radio, April 12, 1994, in FBIS, April 13.

25. John Wallach, to author, May 25, 1994.

26. Ibid. Rolf Ekeus also told this author, on May 25, 1995, that when he had visited Iraq in the spring of 1994, Tariq Aziz had raised the issue of Baghdad's offer to share information about the World Trade Center bombing, and had complained that the United States had not responded.

27. *Chicago Sun Times*, March 5, 1993.

28. Jim Fox, to author, December 1, 1994.

29. *Al Akhbar* and *al Ahram*, March 25, 1993.

30. *New York Times*, August 4, 1997.

31. Ghanim al-Najjar, of Kuwait University, to author, February 23, 1996.

CHAPTER 10: BILL CLINTON, AMERICA, AND SADDAM HUSSEIN

1. *New York Times*, September 23, 1990.

2. Testimony of Ambassador Rolf Ekeus, Senate Permanent Subcommittee on Investigations, Committee on Governmental Affairs, March 20, 1996.

3. In one 1994 poll, 73 percent of Americans said that the United States should have continued fighting until Saddam was overthrown. CBS News/*New York Times*, November 1994, in *National Journal*, December 3, 1994.

4. Baghdad Radio, November 4, 1992, in FBIS, November 5.

5. That is essentially the government's chronology of the bombing

plot. P. 8364 ff., United States v. Muhammad Salameh et al. It is also supported by Nidal Ayyad, who told authorities that Yousef had had money wired from abroad into Salameh's account on November 18, 1992, and that sometime around November 1992 Yousef purchased electronic materials. FD-302, of Nidal Ayyad, July 8, 1993.

6. "Clinton Backs Bush on Iraq but Offers Olive Branch," *New York Times*, January 14, 1993.

7. *Wall Street Journal*/NBC News poll, *Wall Street Journal*, January 29, 1993. Eighteen months later, as Clinton deployed U.S. forces to Kuwait to protect it from possible Iraqi attack, 74 percent of Americans again approved, according to a CNN poll.

8. *Washington Post*, March 30, 1993; *Wall Street Journal*, April 5, 1993.

9. Address by Martin Indyk to the Washington Institute for Near East Policy, Washington, D.C., May 18, 1993.

10. Abd al Jabbar Muhsin, "Saddam Hussein and George Bush," *al-Jumhuriyah*, August 19, 1992.

11. Iraq Radio, February 14, 1993, in FBIS, February 15.

12. *New York Times*, October 25, 1992.

13. To cite my own misconception, see Laurie Mylroie, *Iraq: Options for U.S. Policy* (Washington, D.C.: Washington Institute for Near East Policy, May 1993), p. 11.

14. Salah Mukhtar, *al-Jumhuriyah*, July 3, 1993.

15. Iraq Radio, January 16, 1994, in FBIS, January 17.

16. Nuri Najm al-Marsoumi, "Beware of the Patient Man's Rage," *Babil*, January 20, 1994.

17. Iraq Radio, March 13, 1994, in FBIS, March 14.

18. Iraqi News Agency, March 15, 1994, in FBIS, March 16.

19. Iraq Radio, July 17, 1994, in FBIS, July 18.

20. UNSCOM spokesman Tim Trevan, to author.

21. *Babil*, September 29, 1994.

22. *Al-Jumhuriyah*, October 1, 1994; *al-Jumhuriyah*, October 4, 1994; *al-Jumhuriyah*, October 8, 1994. These threats, and many more, are contained in a publication compiled and produced by FBIS entitled "Iraq: Crisis of October 1994," November 21, 1994.

23. *Washington Times*, November 20, 1994.

24. Jim Hoagland, "Saddam Won't Just Fight the Same War Over Again," *Washington Post*, October 12, 1994.

25. Salah Mukhtar, *al-Jumhuriyah*, January 11, 1995.

26. Tim Trevan, *Saddam's Secrets: The Hunt for Iraq's Hidden Weapons* (London: HarperCollins, 1999), p. 287.

27. This is detailed in ibid., chapter 17, and in "How Iraq's Biological Weapons Program Came to Light," *New York Times*, February 26, 1998.

28. Abu Sirhan [pseudonym for Saddam or Uday], "The Snow Ball," *Babil*, March 15, 1995.

29. Trevan, *Saddam's Secrets*, p. 319.

30. Nuri Najm al-Marsoumi, "A U.S. Misunderstanding which Must Be Removed!" *al-Iraq*, April 11, 1995.

31. UNSCOM report, S/1995/494, June 20, 1995, p. 10.

32. Rolf Ekeus to author, August 8, 1995, before Kamil's defection that day became known. I advised Ekeus that the Iraqis were serious. He remarked, "Funny, you're the only one [who thinks so]." But he was sympathetic to that view, and he reported the Iraqi ultimatum to the Security Council on August 10.

33. *Washington Times*, August 21, 1995.

34. *Al-Ray* (Jordan), November 27, 1997.

CHAPTER 11: DEFECTION OF HUSSEIN KAMIL

1. Scott Ritter, *Endgame: Solving the Iraq Problem—Once and For All* (New York: Simon & Schuster, 1999), p. 105.

2. Frederick Forsyth, *The Fist of God* (New York: Bantam Books, 1994), pp. 94, 368.

3. Uriel Dann, "Getting Even," *New Republic*, June 3, 1991. Indeed, the *Washington Post*'s Jim Hoagland was among the first Westerners to recognize Saddam's demonic qualities. Hoagland interviewed Saddam in the spring of 1975, after Saddam had reached a deal with the shah of Iran over their disputed border that allowed Iraq to crush the Kurdish rebellion then, and he described that interview in "Saddam, Iraq's Suave Strong Man," *Washington Post*, May 11, 1975. Much later, Hoagland recalled to this author how Saddam sat behind a "huge desk," looking at him through "hooded eyes." And he spoke of the Kurds as nonhumans, as if Hoagland "had taken up for a dog he had kicked." Jim Hoagland, to author, November 27, 1999.

4. Gary Milhollin, "Could Iraq Have the Atomic Bomb?" *New York Times*, November 19, 1997.

5. Ibid.

6. *The New Yorker*, April 5, 1999.

7. Public Broadcasting Service's *Frontline*, "Spying on Saddam: Investigating the UN's Dramatic Thwarted Effort to Uncover Iraq's Chemical, Biological and Nuclear Weapons," April 27, 1999.

8. *Sunday Times* (London), February 19, 1995.

9. Rolf Ekeus, to author, December 3, 1999.

10. UNSCOM report to UN Security Council, June 3, 1998.

11. George Robertson, press conference, December 19, 1998.

12. *Chicago Tribune*, January 31, 1999.

13. Ritter, *Endgame,* p. 219.

14. Defense Department Fact Sheet, "Anthrax Is Preferred Biological Warfare Agent," USIS, February 20, 1998.

15. Lieutenant Colonel Edward Eitzen, M.D., M.P.H., chief of the Preventive and Operational Medicine Department at the U.S. Army's biological defense laboratory at Fort Detrick, Maryland, cited a 1970 World Health Organization study entitled "Hypothetical Dissemination by Airplane of 50 Kilograms of Agent along a 2-kilometer Line Upwind of a Population Center of 500,000 People." Testimony before the Permanent Subcommittee on Investigations, Committee on Governmental Affairs, United States Senate, October 31, 1995.

Another study estimated that 100 kilograms of anthrax, disseminated by an airplane flying upwind of Washington, D.C., on a clear, calm night, could kill between 1 and 3 million people. U.S. Congress, Office of Technology Assessment, *Proliferation of Weapons of Mass Destruction: Assessing the Risks*, OTA-ISC-359 (Washington, D.C.: U.S. Government Printing Office, August 1993), p. 54.

16. Eitzen, "Hypothetical Dissemination." As the Defense Department Fact Sheet explained, "Almost all cases of inhalational anthrax, in which treatment was begun after patients have exhibited symptoms, have resulted in death, regardless of post-exposure treatment."

A less grim, but still quite serious picture has been presented by a number of scientists, based on an incident in the Soviet Union. There was an accidental release of anthrax in the town of Sverdlosk in 1979. The deaths that ensued occurred between two and forty-three days after the accident, and only one-quarter of the affected population became ill in the first week after the attack. Thus, it is argued, after a terrorist attack, there would still be time to treat three-quarters of the affected population. Thomas V. Ingelesby, Donald A. Henderson, et al., "Anthrax As a Biological Weapon: Medical and Public Health Management," *Journal of the American Medical Association*, vol. 281. no. 18, May 12, 1999. Critics of that view argue that there is little way to know if the Soviet information is reliable; perhaps there was more than one anthrax release, or some other factor distorted the data.

17. W. Seth Carus, "Biohazard: Assessing the Bioterrorism Threat," *New Republic*, August 2, 1999.

18. U.S. State Department official, to author, March 1996.

19. David Wurmser, *Tyranny's Ally: America's Failure to Defeat Saddam Hussein* (Washington, D.C.: AEI Press, 1999), p. 21.

20. "Iraq Attack," *Insight*, May 13, 1996.

21. Wurmser, *Tyranny's Ally,* p. 25.

22. Jim Hoagland complained about the Clinton administration's

failure to provide funding for the peace-monitoring force in "Failures in Kurdistan," *Washington Post*, March 31, 1996.

23. A. M. Rosenthal, "Saddam Moves Along," *New York Times*, September 6, 1996.

24. Told to author by UNSCOM official, September 16, 1996.

25. Jim Hoagland, "Saddam Prevailed," *Washington Post*, September 29, 1996.

26. *Los Angeles Times,* December 9, 1996.

CHAPTER 12: EXPULSION OF UNSCOM

1. Rolf Ekeus, presentation to the Council on Foreign Relations, Washington, D.C., June 17, 1997.

2. Timothy McCarthy, UNSCOM inspector, to author, June 18, 1997.

3. *The New Yorker*, April 5, 1999.

4. State Department official, to author, June 1996; UNSCOM official, to author, June 1996.

5. Rolf Ekeus, to author, April 11, 1997.

6. Iraqi Television Network, June 22, 1997.

7. Agence France-Presse, June 26, 1997.

8. Ibid., July 25, 1997.

9. Reuters, October 23, 1997.

10. Iraq Television Network, October 24, 1997

11. Ibid., October 27, 1997.

12. Ibid., October 29, 1997.

13. Associated Press, November 2, 1997.

14. *New York Times*, November 10, 1997.

15. Ibid.

16. http://www.pub.whitehouse.gov/uri-res/I2R?urn:pdi://oma.eop. gov.us/1997/1 1 /12/4.text.1. Veterans Day Remarks by the President, Arlington National Cemetery, November 11, 1997.

17. http://www.pub.whitehouse.gov/uri-res/I2R?urn:pdi://oma.eop. gov.us/1997/1 1 /20/9.text.1. Remarks by the President, Cessna Campus Building, Wichita, Kansas, November 17, 1997.

18. *This Week,* ABC News, November 16, 1997.

19. Iraq Television Network, November 16, 1997.

20. Joint Statement by the Permanent Five, November 20, 1997.

21. http://www.pub.whitehouse.gov/uri-res/I2R?urn:pdi://oma.eop. gov.us/1997/1 1/24/7.text.1. Press briefing by National Security Adviser Sandy Berger, White House, November 20, 1997.

22. Iraq News Agency, November 27, 1997.

23. Iraq Television Network, January 17, 1998.

24. *New York Times*, January 25, 1998.

25. Ibid., January 26, 1998.

26. http://www.pub.whitehouse.gov/uri-res/I2R?urn:pdi://oma.eop. gov.us/1998/2 / 9/6.text.1. Joint Press Conference of President Bill Clinton and Prime Minister Tony Blair of Great Britain, White House, February 6, 1998.

27. http://www.pub.whitehouse.gov/uri-res/I2R?urn:pdi://oma.eop. gov.us/1998/2 / 18/5.text.1. Remarks by President Clinton, on Iraq, to Pentagon Personnel, February 17, 1998.

28. *This Week*, ABC News, February 22, 1998.

29. http://www.usembassy.ro/USIS/Washington-File/300/98-02-18/ eur311.htm. Town Hall Meeting, Ohio State University, Columbus, Ohio, February 18, 1998.

30. *Washington Post*, March 1, 1998.

31. Ibid.

32. Ibid. Scott Ritter, *Endgame: Solving the Iraq Problem, Once and For All* (New York: Simon & Schuster, 1999), p. 180, offers a different version. According to Ritter, Albright told Annan that the administration would back any agreement that would return UNSCOM to Iraq.

33. Kofi Annan, Press Conference, UN Headquarters, February 25, 1998.

34. http://www.pub.whitehouse.gov/uri-res/I2R?urn:pdi://oma.eop. gov.us/1998/2 / 24/3.text.1. Remarks by President Clinton on Iraq, White House, February 23, 1998.

35. *Washington Post*, February 26, 1998.

36. Ritter, *Endgame,* p. 182.

37. *Washington Post*, March 1, 1998.

38. Reuters, February 9 and February 17, 1998.

39. *Al-Quds al-Arabi*, February 23, 1998.

40. Samuel Berger, "Iraq: Securing America's Interests," *Washington Post*, March 1, 1998.

41. Opening statement of Secretary of State Madeleine K. Albright, before the Subcommittee on Foreign Operations, Export Financing and Related Programs, House Appropriations Committee, March 4, 1998.

42. Ritter, *Endgame,* p. 185; *Washington Post*, October 12, 1998.

43. Ritter, *Endgame,* p. 185.

44. So Ritter suggested, ibid., p. 182.

45. Richard Butler, *Talk*, September 1999.

46. Ibid.

47. In addition to her trying to pull Ritter off the UNSCOM team when inspections resumed in March, it was also revealed that on August 4, Albright advised Butler not to proceed with an inspection

planned for August 6. Barton Gellman, "U.S. Fought Surprise Iraqi Arms Inspections: Visits Canceled after Albright Argued Timing Wrong for Confrontation," *Washington Post*, August 14, 1998.

48. "Illusory Inspections in Iraq," *New York Times*, August 28, 1998; "Partners in a Charade" and "Saddam Wins," *Washington Post*, August 15 and 28, 1998; "Iraq Policy Meltdown," *Washington Times*, August 17, 1998; and Martin Peretz, "UNSCOMSCAM," *New Republic*, September 7, 1998.

49. Senate Foreign Relations Committee, Subcommittee on Near Eastern and South Asian Affairs, September 9, 1998.

50. Ibid.

51. Jim Hoagland, "No Time to Tone Down," *Washington Post*, April 23, 1998.

52. Joint Hearing of Senate Foreign Relations Committee and Energy and Natural Resources, May 21, 1998.

53. Reuters, May 6, 1998.

54. Iraq Television Network, May 1, 1998.

55. http://www.pub.whitehouse.gov/uri-res/I2R?urn:pdi://oma.eop. gov.us/1998/1 1 /16/4.text.1. Veterans Day Remarks by the President, Arlington National Cemetery, November 11, 1998.

56. *Washington Post*, November 17, 1998.

57. http://www.pub.whitehouse.gov/uri-res/I2R?urn:pdi://oma.eop. gov.us/1998/1 1 /15/1.text.1. Press Briefing by the President, White House, November 15, 1998.

58. Reuters, November 19, 1998.

59. "This Time We Mean It, Maybe," *Washington Post*, November 24, 1998.

60. *Washington Post*, December 7, 1998.

61. *New York Times*, December 7, 1998.

62. Ibid., December 16, 1998.

63. http://www.pub.whitehouse.gov/uri-res/I2R?urn:pdi://oma.eop. gov.us/1998/1 2 /19/5.text.1. Press Briefing by the President on Iraq, White House, December 16, 1998.

64. http://www.defenselink.mil/news/Dec1998/t12171998t217dfox. html. Press Briefing by Secretary of Defense William Cohen, Pentagon, December 17, 1998.

65. http://secretary.state.gov/www/statements/1998/981217.html. Press Briefing by Secretary of State Madeleine Albright, State Department, December 17, 1998.

66. Jim Lehrer News Hour, December 21, 1998.

67. "Wag the Fox," *New Republic*, January 4, 1999.

68. A. M. Rosenthal, "What Clinton Can Do," *New York Times*, December 18, 1998.

69. *New York Times*, August 25, 1999.

CHAPTER 13: OTHER LINKS

1. Jim Fox, to ABC News, as told to the author by Sheila MacVicar during ABC/*Newsweek* investigation, April–June 1994. Fox subsequently repeated that view to the author on several occasions.

2. Government Exhibit 810, United States v. Muhammad Salameh et al., is the phone record from the apartment in which Nidal Ayyad lived in 1992. It shows calls from Ayyad's apartment to Yasin's number as listed in the 1992 Jersey City phone directory through May 16, 1992, after which the number ceases to appear in Ayyad's phone records.

3. *New York Newsday*, "Jersey City Connection," June 17, 1993.

4. Ibid.

5. Ibid.

6. Government Exhibit 585-A, Ryder Rental Agreement with Muhammad Salameh, United States v. Muhammad Salameh et al..

7. Government Exhibt 657, Admission Records for Ramzi Yousef and Muhammad Salameh, Rahway Hospital, United States v. Muhammad Salameh et al.

8. Report by Dr. Russoniello, January 24, 1993.

9. *New York Times*, May 26, 1993.

10. "Jersey City Connection," *New York Newsday,* June 17, 1993. *Newsday* reporter Kevin McCoy, who contributed to that story, told this author that according to motor vehicle records, it was shortly before the accident that Musab Yasin transferred the title of his car to New York in the names of Yousef and Salameh.

Musab Yasin himself confirmed that he had registered his car under Yousef's and Salameh's names. He claimed that was because he "had problems concerning the license." Jamal Khashoggi, "He Speaks to *al Hayat* about His Relations with Muhammad Salameh and Ramzi Yousef; Musab Yasin: 'I Was Exposed to American Pressure to Involve Iraq in the Trade Center Bombing,' " *al-Hayat*, January 25, 1994. Yasin told *al-Hayat* that he was a Muslim fundamentalist but stayed away from its more extreme manifestations. But as *al Hayat* noted, "This does not really explain his good relation with Ramzi Yousef."

11. Sworn statement by Musab Yasin, June 8, 1993.

12. Official at the Office of the District Attorney of New York, to author. The analysis of the telephone records was given to the official by this author on January 13, 1995. We discussed the matter several times subsequently, until June, when he reached the conclusion that he could do nothing.

13. P. 3604, transcript of United States v. Ramzi Yousef et al. [airplane bombing conspiracy]; *Los Angeles Times*, May 30, 1995.

14. Letter from Federal Aviation Administration to investigative

journalist Brad White in response to a Freedom of Information Act request, March 6, 1996.

15. FBI summary of Yousef's statement, provided to the author by *Daily News* reporter Greg Smith, p. 1.

16. Jim Fox, to author, August 16, 1995.

17. *New York Times*, August 4, 1995.

18. *Dallas Morning News*, June 8, 1997.

19. *New York Times*, August 6, 1997.

20. FBI summary of Yousef's statement, provided to the author by *Daily News* reporter Greg Smith, p. 2.

21. Ibid., p. 3.

22. *Dallas Morning News,* June 8, 1997.

23. Government Exhibit 221, United States v. Ramzi Yousef et al. [Trade Center bombing].

24. Ibid.; *Dallas Morning News*, November 4, 1997.

25. Government Exhibit 221, United States v. Ramzi Yousef et al. [Trade Center bombing].

26. Ibid.

27. Ibid.

28. Ibid.; *New York Times,* November 13, 1997.

29. Indeed, that was the general view in New York. Sheila MacVicar, narrating the ABC News program *Day One* on June 24, 1994, described the Abdul Basit passport as "still another assumed name and nationality." Jim Fox said on the program, "Yousef is really a shadowy figure in this whole thing. There is so little that we know about him for certain." *Newsweek*, July 4, 1994, described Abdul Basit Karim as "yet another alias."

Similarly, a book on the Trade Center bombing written by a team of *Newsday* reporters—Jim Dwyer et al., *Two Seconds under the World* (New York: Crown Publishing, 1994), p. 62—dismissed the Abdul Basit identity as yet one more alias that Yousef had used.

CHAPTER 14: ATTEMPTED BOMBING OF THE UN

1. *New York Newsday*, October 28, 1993.

2. Cited in *New York Times*, August 28, 1993.

3. Ibid.

4. Government's Memorandum of Law in Opposition to Defendants' Pretrial Motions (Phase 1), p. 41, United States v. Omar Abdel Rahman et al.

5. Response by Office of the United States Attorney, Southern Dis-

trict of New York, to Section C: The World Trade Center Bombing; February 27, 1997, p. 2.

6. Jim Fox, to author, March 21, 1996.

7. In discussing Mubarak's assassination, the conspirators referred to him as "the laughing cow." The U.S. government gave this phrase a particularly sinister interpretation. Of course, the discussion of Mubarak's assassination was indeed quite sinister. But "The Laughing Cow" (*La vache qui rit*) is a soft, bland French cheese, popular in Egypt. The cheese comes in a little round box sporting an illustration of a smiling cow, and it is popularly thought that the cow bears a resemblance to Mubarak. Egyptians are known among Arabs for their sense of humor. Referring to Mubarak as "*La vache qui rit*" was, for many, a way of making fun of him and alluding to what was considered his lack of substance.

Egyptians tell another joke to that effect, playing on the fact that Mubarak never chose a vice president. It goes as follows: Nasser looked around for someone more stupid than himself to be vice president, and he chose Sadat. Sadat, when he became president, looked around for someone more stupid than himself, and he chose Mubarak. And Mubarak, when he became president, looked around for someone more stupid than himself, and. . . .

8. Government's Memorandum of Law in Opposition to Defendants' Pretrial Motions (Phase 1), p. 30, United States v. Omar Abdel Rahman et al.

9. Ibid., pp. 35, 36.

10. Ibid.

11. *New York Newsday*, May 16, 1994.

12. Government's Memorandum of Law in Opposition to Defendants' Pretrial Motions (Phase 1), p. 38, United States v. Omar Abdel Rahman et al.

13. Ibid., p. 48.

14. Bernard Lewis, America's leading scholar of Islam, has explained this dissonance in many works, including *Islam and the West* (New York: Oxford University, 1993); cited in Table of Authorities, Government's Memorandum of Law in Opposition to Defendants' Pretrial Motions (Phase 1), United States v. Omar Abdel Rahman et al.

However, the U.S. government pointed to widely held Muslim beliefs and portrayed them as an integral part of the New York bombing conspiracies. For example, the prosecution claimed, "Among the tenets of the defendants' organization is the belief that the only valid law has been revealed by Allah through the Quran and the other authoritative writings of Islam." That is a traditional Islamic view, and the overwhelming majority of those who hold it are not terrorists.

An equivalently offensive misrepresentation of Catholics would be to attribute murder and mayhem at abortion clinics to the church's opposition to abortion. An equivalent for Jews would be to attribute Baruch Goldstein's February 1994 slaughter of Muslims as they prayed in a Hebron mosque to the belief that God made a covenant with the Jews and gave the land of Israel to them.

15. Nosair also opposed bombing the UN. When Salem and Siddig Ali visited Nosair at Attica Penitentiary on May 20, 1993, Nosair advised against any bombings for the present. He wanted to get out of prison, and he had concluded that a bombing campaign would not free him. Instead, Nosair suggested kidnapping Richard Nixon and Henry Kissinger and holding them hostage in exchange for his freedom and that of the Trade Center bombing defendants.

16. Vincent Cannistraro, to author, October 12, 1994.

17. December 20, 1995, p. 28, United States v. Omar Abdel Rahman et al.

18. *New York Times*, April 11, 1996.

19. Ibid.

20. *New York Times*, June 28, 1993.

21. Retired senior official in state counterterrorism, to author, December 13, 1999.

22. *New York Times*, August 26, 1998.

23. Tim Trevan, former UNSCOM spokesman, to author, October 1999.

24. Al-Ta'ish holds the rank of *Mudir Amm*, or general director, in the Iraqi *Mukhabarat*. Since the late 1980s he has been assigned to the Foreign Ministry, under whose auspices he worked in Mauritania and Morocco before arriving in Sudan.

25. *Babil*, June 26, 1993, in FBIS, June 27.

26. *New York Times*, June 28, 1993.

27. And Baghdad did get the message. In *al-Jumhuriya* on July 3, a prominent Iraqi journalist, Salah Mukhtar, wrote, "It is a wrong conclusion on Clinton's part that striking the intelligence headquarters will ease the Iraqis' anger. This only confirms Clinton's weakness and naïveté." In other words, Saddam had no intention of permanently ending his terror. To be sure, there were no more major terrorist assaults on U.S. targets for a year and a half—until Ramzi Yousef's second bombing plot emerged.

28. In addition, the author held a series of discussions from May 1993 to December 1994 with Martin Indyk, NSC adviser on the Middle East. In the first of those discussions, Indyk seemed unaware of the suspicions that Iraq was behind the Trade Center bombing. But by the last of them, he did seem to know about such suspicions and about the belief

of his superiors at the White House that they had taken care of the terrorism in New York by the strike on Iraqi intelligence headquarters.

29. Government's Memorandum of Law in Opposition to Defendants' Pretrial Motions (Phase 1), p. 47, United States v. Omar Abdel Rahman et al.

CHAPTER 15: RAMZI YOUSEF'S SECOND CONSPIRACY

1. State Department official, to author, February 1995, shortly after Yousef's arrest.

2. *New York Times*, February 15 and 21, 1995.

3. *The Nation* [Islamabad], February 10, 1995.

4. Indictment of Ramzi Yousef et al., S12 93 Cr. 180 (KTD), p. 25; P. 4103, United States v. Ramzi Yousef et al. [airplane bombing conspiracy].

5. *New York Times*, February 12, 1995.

6. *Al Hayat*, April 12, 1995.

7. Raghida Dirgham, to author, May 30, 1995.

8. P. 2674, United States v. Ramzi Yousef et al. [airplane bombing conspiracy].

9. Ibid., p. 883.

10. State Department official, to author, May 1996.

11. P. 4097, United States v. Ramzi Yousef et al. [airplane bombing conspiracy].

12. Ibid., pp. 4165–66.

CHAPTER 16: OFFICIAL MISINFORMATION

1. Justice Department official, to author, January 31, 1996.

2. Mary Anne Weaver, an American journalist who has long followed the Islamic extremists, particularly in Egypt, gives a different account of Ramzi Yousef's identity in "Children of the Jihad," *The New Yorker*, June 12, 1995, and in *A Portrait of Egypt: A Journey through the World of Militant Islam* (New York: Farrar Straus and Giroux, 1999). Her information comes from unnamed Pakistani intelligence sources. According to those sources, Yousef's real identity is Abdul Basit, and he and his family left Kuwait for Turbat, in Baluchistan, in 1986 and resided there subsequently. Thus, Weaver writes, "In the fall of 1987, Abdul Basit left Turbat to enroll in an electrical engineering course at the West Glamorgan Institute of Higher Education."

That would have happened, of course, if the Pakistani information were correct. But the Pakistanis had reason at that time to disseminate information, regardless of whether it was true or not. Pakistan has a dubious record on terrorism. And in March 1995, two Americans were killed in a revenge attack following Yousef's arrest. Subsequently, Pakistani authorities released a lot of information to journalists that implied they were knowledgeable about their terrorism problem and were taking effective measures to deal with it. That information included a long, typewritten account of the investigation of an individual referred to as "Abdul Shakoor," who was said to be an associate of Ramzi Yousef. According to the widely disseminated report, Yousef and Abdul Shakoor were involved in an attempt to kill Pakistani prime minister Benazir Bhutto, and Yousef was also involved in the bombing of a Shi'i shrine in Iran. The information the Pakistanis put out included the claim that Abdul Basit and his family had moved from Kuwait to Baluchistan in 1986, first reported by Weaver in her *New Yorker* article.

But that is contradicted by several other sources. Abdul Basit's teachers remembered that Abdul Basit lived in Kuwait when he was a student at Swansea. His passports show him traveling between Kuwait and Great Britain in that period, not between Pakistan and Great Britain. Finally, Kuwait's Interior Ministry file on Abdul Basit has the family living in Kuwait until Iraq's invasion. Although the latter two sets of documents are not fully reliable, it is hard to understand why anyone would have tampered with them to make it appear that Abdul Basit and his family were resident in Kuwait when in fact they lived in Baluchistan.

And crediting the Pakistani information produces a composite person—Ramzi Yousef-cum-Abdul Basit. Thus, Weaver writes that "as a teen-ager" Abdul Basit was "tall and handsome." Of course, Abdul Basit was not tall; Ramzi Yousef was.

As for the Pakistani claim that Yousef participated in an attempt on Bhutto's life and the bombing of a Shi'i shrine in Iran, that information has not been independently confirmed either. And there is reason to doubt it. Significantly, there was never a trial in Pakistan for Abdul Shakoor and the other individuals named in that report, as there should have been, if they had really conspired to kill the prime minister.

Simon Reeve, in *The New Jackals: Ramzi Yousef, Osama bin Laden, and the Future of Terrorism* (Boston: Northeastern University Press, 1999), similarly accepts the Pakistani information unquestioningly, including the claim of Yousef's role in the other terrorism alleged by the Pakistanis and the claim that Abdul Basit and his family left Kuwait for Baluchistan in 1986 and resided there subsequently. Reeve departs from Weaver, however, when he claims that Yousef was back in Kuwait when Iraq invaded in August 1990, and that he then assisted the Iraqi occupation forces.

3. Reeve, *The New Jackals*, p. 251.

4. After misstating Abdul Basit Karim's height, Reeve writes, "Neil Herman (recently retired as FBI Supervisory Special Agent in charge of New York's Joint Terrorist Task Force) and the FBI are convinced Yousef and Karim are one and the same" (p. 251).

After reading that, I spoke with Herman, on February 22, 2000. Herman told me that he did *not* believe that Yousef's real identity was Abdul Basit. In Herman's view, Abdul Basit was just one of many aliases Yousef had used. By saying that Yousef and Karim were one and the same, Herman meant that the individual who left the United States the night of the Trade Center bombing as Abdul Basit was, in fact, Yousef.

Indeed, earlier in the same chapter, in a contradiction Reeve apparently failed to notice, he quoted Judge Duffy who, addressing Yousef at his sentencing hearing, said, "We don't even know what your real name is" (p. 243).

5. Conversation held in April 1995.

6. Conversation held in April 1995.

7. An investigation like the Trade Center bombing is very difficult. And it produces an enormous amount of information. Moreover, the primary task of New York law enforcement was to identify the individuals responsible, arrest them, and succeed in gaining convictions. Thus, it can happen that an important point, particularly one not related to that task, could be overlooked. And that, in fact, did happen. Fox and others knew that Yousef was not Abdul Basit, but they did not take the further step of asking, If Yousef is not Abdul Basit, then how did his fingerprints get into a file in Kuwait?

8. Indictment of Ramzi Yousef et al., S7 94 CR.180 (KTD), p. 16.

9. James C. McKinley, Jr., "Bomb Suspect Says the U.S. Merits Attack," *New York Times*, March 25, 1995. McKinley told me (March 29, 1995) that only one source—in the U.S. attorney's office—claimed to know Yousef's real identity—that is, Abdul Basit. But since that source would speak only off the record, his reporter's instincts told him not to use it.

10. Conversation held on February 14, 1995.

11. Conversation held on March 17, 1995.

12. See indictments of Ramzi Yousef et al., S4 93 CR.180 (KTD) through S7 94 CR.180 (KTD).

13. Jim Fox, to author, February 1995.

14. *New York Times*, October 30, 1993.

15. James C. McKinley, Jr., "Question in Bombing Trials: Can Two Masterminds Exist?" *New York Times*, April 15, 1995.

CHAPTER 17: THE ISLAMIC CHANGE MOVEMENT

1. "Extremist Saudi Organization Threatens to Hit Foreigners and Influential Princes in the Kingdom," *al-Quds al-Arabi*, April 11, 1995.

2. "Unknown Saudi Movement Hints at Carrying Out Its Threats by Hitting Foreign Forces," *al-Quds al-Arabi*, July 3, 1995.

3. Iraq Television, November 14, 1995.

4. Reuters, November 14, 1995.

5. *Al-Jumhuriyah*, November 14, 1995.

6. Agence France-Presse, November 16, 1995, citing *al-Yawm*.

7. *New York Times*, November 16, 1995.

8. Uri Dan and Dennis Eisenberg, "Sly Snake of Baghdad," *Jerusalem Post*, November 23, 1995.

9. Ibid.

10. Reuters, November 15, 1995.

11. *Washington Times*, November 17, 1995.

12. http://www.pub.whitehouse.gov/uri-res/I2R?urn:pdi://oma.eop.gov.us/1995/11/13 / 18.text.1. The White House, Office of the Press Secretary, "Statement by the President," November 13, 1995.

13. http://www.pub.whitehouse.gov/uri-res/I2R?urn:pdi://oma.eop.gov.us/1995/11/13 / 13.text.1 Remarks by the President, White House, November 13, 1995.

14. *Wall Street Journal*, November 14, 1995.

15. *Washington Times*, November 29, 1995.

16. Professor Ghanim al-Najjar, of Kuwait University, to the author. He had participated in the meeting of a delegation from the Arab Political Science Association with Tariq Aziz.

17. Thomas L. Friedman, "Before It's Too Late," *New York Times*, December 10, 1995.

18. *New York Times*, June 27 and 29, 1996.

19. As the *New York Times*, July 7, 1996, reported, "The CIA and other government experts . . . significantly misjudged the bomb-making capabilities of militants in Saudi Arabia, concluding that they could not build a bomb larger than the 200-pound bomb that killed five Americans . . . in Riyadh last November."

20. Reuters, June 24, 1996.

21. Iraq News Agency, June 25, 1996.

22. Ibid.

23. Uri Dan and Dennis Eisenberg, "Saddam's Sly Revenge," *Jerusalem Post*, June 27, 1996.

24. Uri Dan and Dennis Eisenberg, "Terrorists' Haven," *Jerusalem Post*, July 11, 1996.

25. *Sunday Times* (London), February 19, 1995.

26. *The Independent*, July 5, 1996.

27. Richard Perle, "At Long Last, America Takes the Gloves Off," *Sunday Times* (London), August 23, 1998.

28. Iraqi News Agency, June 27, 1996.

29. Mother of Battles Radio, June 27, 1996.

30. *MidEast Mirror*, July 10, 1996.

31. http://www.pub.whitehouse.gov/uri-res/I2R?urn:pdi://oma.eop. gov.us/1996/6 / 26/11.text.1. Remarks by the President, White House, June 25, 1996.

32. http://www.pub.whitehouse.gov/uri-res/I2R?urn:pdi://oma.eop. gov.us/1996/6 / 27/4.text.1. Remarks by the President, Perouges, France, June 27, 1996.

33. *Washington Times*, June 27, 1996.

34. *Washington Post*, July 3, 1996.

35. *Washington Post*, July 8, 1996.

36. *New York Times*, August 3, 1996.

37. One scholar argued that Iran, rather than Iraq, was behind the bomb, on the explicit assumption that the 1991 Gulf War had been a "complete Iraqi defeat." Ely Karmon, "Tehran Starts and Stops," *Middle East Quarterly*, December 1998.

38. Bill Gertz, "Saudi Financier Tied to Attacks," *Washington Times*, October 24, 1996.

That information, in its essentials, was included in the indictment of bin Ladin after the bombings of the U.S. embassies in Kenya and Tanzania. The indictment did not explicitly charge bin Ladin with either of the Saudi bombings, but it stated that "on at least two occasions," between 1992 and 1995, "members of al Qaeda," bin Ladin's organization, "transported weapons and explosives from Khartoum in the Sudan to the coastal city of Port Sudan for trans-shipment to the Saudi Arabian peninsula." Indictment S (6) 98 Cr. 1023 (LBS), p. 18.

39. Indeed, the al-Khobar bombing produced a rash of articles in the U.S. media questioning the stability of the Saudi government; see, for example, "The Saudis: Bombing Attack Raises Questions about Saudi Government's Stability," *New York Times*, June 27, 1996; "U.S. Fears Bombing Reveals New Cracks in Saudi Society," *New York Times*, June 30, 1996; "Saudis' Heartland is Seething with Rage at Rulers and U.S.," *New York Times*, November 5, 1996.

40. "Saudis Hold Forty Suspects in G.I. Quarters Bombing," *Washington Post*, November 1, 1996.

41. *Jerusalem Post*, January 21, 1997.

42. Thomas L. Friedman, "Where Is Jaafar?" *New York Times*, January 8, 1997.

43. *Washington Post*, January 23, 1997.

44. *Washington Post*, June 28, 1997.

45. Indeed, this author was present during a discussion, March 21, 2000, between a former CIA official and a prominent Iraqi opposition figure, when the question of the al-Khobar bombing arose. The former CIA official maintained that the mere fact that certain individuals left Saudi Arabia for Iran after the bombing indicated that they were involved in the attack and that Iran was behind it. The other man suggested that those individuals, involved in illegal political activity, might well have left because, given the circumstances, they had reason to fear they would be arrested, and that their mere leaving did *not* tie them to the bombing. The former CIA official conceded that this might well have been so.

46. "Damascus Declaration Foreign Ministers Meet in Cairo," Saudi Press Agency, December 29, 1996.

47. *Jerusalem Post*, January 7, 1997.

48. *Jerusalem Post*, January 9, 1997.

49. Uri Dan and Dennis Eisenberg, *Jerusalem Post*, November 23, 1995, June 27, 1996, July 11, 1996, and January 7, 1997.

Chapter 18: Other Terrorist Conspiracies

1. Indictment of Usama bin Ladin et al., S (6) 98 CR.1023 (LBS), p. 30.

2. Ibid.

3. Text of World Islamic Front's Statement Urging Jihad against Jews and Crusaders, *al-Quds al-Arabi*, February 23, 1998.

4. Iraq Television, May 1, 1998.

5. Indictment of Usama bin Ladin et al., S (6) 98 CR. 1023 (LBS), p. 32.

6. Ibid., p. 30.

7. Iraq News Agency, May 17, 1998.

8. Indictment of Usama bin Ladin et al., S (6) 98 CR. 1023 (LBS), p. 32.

9. Ibid., p. 31.

10. Ibid., p. 32.

11. Iraq Satellite Television, July 17, 1998.

12. Iraq Radio, July 21, 1998.

13. Indictment of Usama bin Ladin et al., S (6) 98 CR. 1023 (LBS), p. 36.

14. Ibid.

15. Richard Butler, *Talk*, September 1999.

16. Ibid.

17. Iraq Satellite Television, August 5, 1998.

18. Ibid.

19. Indictment of Usama bin Ladin et al., S (6) 98 CR. 1023 (LBS), p. 37.

20. Ibid.

21. Ibid.

22. http://www.pub.whitehouse.gov/uri-res/I2R?urn:pdi://oma.eop. gov.us/1998/8 / 9/7.text.1. Remarks by the President, White House, August 7, 1998.

23. Ambrose Evans-Pritchard, "Did Saddam Pull the Strings of the Terrorist Bombers?" *Daily Telegraph*, August 12, 1998.

24. Helle Bering, "Did Saddam Send the Bombers?" *Washington Times*, August 13, 1998.

25. John Cooley, *Unholy Wars: Afghanistan, America and International Terrorism* (Sterling, Va.: Pluto Press, 1999), p. 223.

26. *Milan Corriere della Sera* (Internet version), February 1, 1999; *Al-Watan al-Arabi* (Paris), January 1, 1999; Yossef Bodansky, *Bin Laden: The Man Who Declared War on America* (Rocklin, Calif.: Forum, 1999), p. 361.

Newsweek, January 11, 1999, reported that there had been low-level contacts between Iraqi intelligence and al-Qaeda. But it also quoted "a well-informed administration official" as saying, "I'm skeptical that Saddam would resort to terrorism."

27. http://www.pub.whitehouse.gov/uri-res/I2R?urn:pdi://oma.eop. gov.us/1998/8 / 20/5.text.2. Address to the Nation by the President, White House, August 20, 1998.

28. http://www.pub.whitehouse.gov/uri-res/I2R?urn:pdi://oma.eop. gov.us/1998/8 / 21/1.text.1. Press Briefing by Secretary of State Madeleine Albright and National Security Adviser Sandy Berger, White House, August 20, 1998.

29. *New York Times*, October 27, 1999.

30. *New York Times*, December 18, 1999.

31. http://www.pub.whitehouse.gov/uri-res/I2R?urn:pdi://oma.eop. gov.us/1998/8 / 21/1.text.1. Press Briefing by Secretary of State Madeleine Albright and National Security Adviser Sandy Berger, White House, August 20, 1998.

32. *New York Times*, August 25 1998.

33. *New York Times*, October 27, 1999.

34. Reuters, August 25, 1998.

35. *New York Times*, August 26, 1998.

36. *Los Angeles Times*, August 26, 1998.

37. *New York Times*, August 25, 1998.

38. *Washington Post*, September 1, 1998.

39. *New York Times*, October 4, 1998.

40. *Washington Post*, July 25, 1999.
41. *Washington Post*, May 7, 1999.
42. Indictment of Usama bin Ladin et al., S (6) 98 CR. 1023 (LBS), p. 40.

CHAPTER 19: "INTERNATIONAL RADICAL TERRORISM"

1. Daniel Benjamin and Steven Simon, "The New Face of Terrorism," *New York Times*, January 4, 2000.

2. William S. Cohen, "Preparing for a Grave New World," *Washington Post*, July 26, 1999.

3. Terrorist Research and Analytical Center, National Security Division, Federal Bureau of Investigation, *Terrorism in the United States, 1994*, p. 14.

4. Ibid.

5. Ibid.

6. Vincent Cannistraro, to author, April 1996.

7. State Department official, to author, April 1996.

8. Steven Emerson, "A Terrorist Network in America?" *New York Times*, April 7, 1993.

9. Indeed, that is what reviewers of the documentary for two major newspapers understood—no state sponsorship. See " 'Jihad': Radical Theories on PBS," *Washington Post*, November 21, 1994; and "In 'Jihad in America,' Food for Uneasiness," *New York Times*, November 21, 1994.

10. *New York Times*, April 25, 1996. And he wrote similarly the year before in Steven Emerson, "The Other Fundamentalists," *New Republic*, June 12, 1995. He concluded, "As the World Trade Center bombing shows, it does not take a big group to make a big catastrophe."

11. Steven Emerson, "The Danger of Radical Islam," interview with Daniel Pipes, *Middle East Quarterly*, June 1997. Pipes, editor of the *Middle East Quarterly*, is far less strident than Emerson, but he tends to take a similar position. In "America's Muslims against America's Jews," *Commentary*, May 1999, Pipes argues that Muslim extremists in the United States are a much greater danger than American Jews generally recognize. But most of the examples Pipes uses to make that point either do not involve extremists or do not involve extremists alone. The article leads with the description of an aborted pipe bombing intended for a subway stop in a Jewish neighborhood in New York. But neither of the two Palestinians charged in the plot was a Muslim extremist. Similarly, Pipes cites the two 1993 New York bombing conspiracies, which, of course, were not carried out by extremists alone. And Pipes, like Emerson, writes as if those charged in the second New York bombing conspiracy also participated in the Trade Center bombing.

Daniel Pipes, "In Muslim America: A Presence and a Challenge," *National Review*, February 21, 2000.

12. John Cooley's *Unholy Wars: Afghanistan, America, and International Terrorism* (Sterling, Va.: Pluto Press, 1999) is written in the same vein. Yet at least Cooley knew there was an indicted fugitive in Iraq and noted, "This has raised questions by some analysts who suspect the hand of Saddam Hussein in the attack"; p. 237.

13. Simon Reeve, *The New Jackals: Ramzi Yousef, Usama bin Laden, and the Future of Terrorism* (Boston: Northeastern University Press, 1999), pp. 2, 266.

14. Ibid., p. 246. Reeve quotes Jim Fox in this context: " 'Although we are unable to say with certainty the Iraqis were behind the [WTC] bombing, that is the theory accepted by most of the veteran investigators' (James Fox, speaking in 1994, before Yousef's capture)." The quoted language appears, however, to come from a letter, dated October 24, 1994, written by Fox in support of an early draft of this book: "I found it to be extremely accurate, and *although we are unable to say with certainty the Iraqis were behind the bombing, that is certainly the theory accepted by most of the veteran investigators*" (emphasis added).

Fox passed my manuscript on to colleagues in New York, and it thereafter circulated in some official circles in Washington. The essentials of the argument were also presented in Laurie Mylroie, "The World Trade Center Bomb: Who Is Ramzi Yousef? Why It Matters," *National Interest*, Winter 1995–1996. Reeve draws on the following points made in that article to conclude that Iraq was probably behind the Trade Center bomb: the timing of the bombing, on a Gulf War anniversary; the fact that the fugitive, Abdul Rahman Yasin, is currently in Baghdad and working for the Iraqi government; the fact that Salameh called his uncle in Baghdad frequently, in the late spring of 1992, when he had been drawn into Nosair's pipe bombing plot (Reeve, pp. 246, 247).

15. Ibid., pp. 63, 66.

16. Ibid., p. 83.

17. Ibid., p. 246.

18. Indeed, the assertion is contradicted by Yousef's own statement to the FBI agents who accompanied him back to America after his arrest. In it he claimed that both his parents were Pakistani, as discussed in chapter 13.

19. Reeve, p. 127.

20. Reeve reports that the FBI found $2,615 cash in the conspirators' apartments, underscoring just how unnecessary it was for Salameh to return to the Ryder agency for the deposit on the van he had rented, exposing himself to easy arrest (p. 246).

21. The conversation with Neil Herman occurred on February 22, 2000.

The New Jackals was written with startling speed. In November 1998, Reeve posted this message on the Internet: "I'm just starting a research project on Ramzi Yousef—can anyone suggest the best source for information on him? Does anyone know is [sic] the leading expert on Yousef?? Your help would be much appreciated" (http://www.nonviolence. org/board/messages/5040.htm). Although Reeve had not worked previously on the Middle East, the book was in print a year later and included bin Ladin's terrorism as well.

Why would a foreign journalist, virtually unknown in the United States, be given the level of special access to U.S. official sources that is evident throughout Reeve's book? That sort of access is often associated with a particular interpretation, or spin, that is being promoted by certain government agencies. Possibly, the administration wanted to put out its own spin on the terrorist threat, which loomed as a particularly acute problem in the wake of the embassy bombings.

22. Christopher M. Andrew and Vasily Mitrokhin, *The Sword and the Shield: The Mitrokhin Archive and the Secret History of the KGB* (New York: Basic Books, 1999), p. 86.

23. Ibid., p. 379.

24. United States Department of State, *Patterns of Global Terrorism: 1990* (Washington, D.C.: Department of State publications, 1991), p. 8.

25. Report of the Commission on the Roles and Capabilities of the United States Intelligence Community, *Preparing for the Twenty-First Century: An Appraisal of U.S. Intelligence*, March 1, 1996, p. 38.

26. Robert M. Gates, *From the Shadows: The Ultimate Insider's Story of Five Presidents and How They Won the Cold War* (New York: Simon & Schuster, 1996), p. 204.

27. Ibid.

28. Report of the Commission, p. 38.

29. "First Annual Report to the President and the Congress of the Advisory Panel to Assess Domestic Response Capabilities for Terrorism Involving Weapons of Mass Destruction," December 15, 1999, p. 12.

30. Ibid., p. 15.

31. Clinton said, "Think how many can be killed by just a tiny bit of anthrax, and think about how it's not just that Saddam Hussein might put it on a SCUD missile . . . and send it on to some city he wants to destroy. Think about all the other terrorists and other bad actors who could just parade through Baghdad and pick up the stores if we don't take action." Jim Lehrer NewsHour, January 21, 1998.

32. First Annual Report, p. 18.

CHAPTER 20: CONCLUSION

1. Laurie Mylroie, "The World Trade Center Bomb: Who Is Ramzi Yousef; Why It Matters," *The National Interest* (Winter 1995/6).

2. "Scenarios of a Coming War," *Babil,* September 20, 2001. No author is given, but the article is generally assumed to have been written by Uday.

3. Jim Hoagland, "What About Iraq?" *Washington Post,* October 12, 2001. Six days later, R. James Woolsey, referring to Hoagland's comment, wrote, "I checked yesterday and essentially the same situation still obtains." R. James Woolsey, "Know Thy Enemy: The Iraq Connection," *Wall Street Journal,* October 18, 2001.

Index

An "f" following a page number indicates that information is found in a figure, illustration, or chart. An "n" following a page number means that information is found in an endnote. A "t" following a page number indicates that information is found in a table.